KU-761-087

PHYSICS

TEACH YOURSELF BOOKS

PHYSICS

David Bryant

TEACH YOURSELF BOOKS

Hodder and Stoughton

First printed 1971
Second edition 1985
Third impression 1988

Copyright © 1971 edition
Hodder and Stoughton Ltd
Copyright © 1985 edition
David Bryant

British Library Cataloguing in Publication Data

Bryant, D. (David), *1936–*
Physics.–2nd ed.–(Teach yourself books)
1. Physics
I. Title
530   QC21.2

ISBN 0 340 37958 8

Printed in Great Britain
for Hodder and Stoughton Educational,
a division of Hodder and Stoughton Ltd,
Mill Road, Dunton Green, Sevenoaks, Kent,
by Richard Clay Ltd, Bungay, Suffolk
Photo Typeset by Macmillan India Ltd, Bangalore

Available in the USA from Random House, Inc.,
201 East 50th Street, New York, NY 10022

ISBN 0 679 10406 2 in USA

Contents

Acknowledgments

Several industrial concerns have been kind enough to supply photographs for this book and other authors and publishers have allowed me to reproduce their illustrations. To all of them I offer my thanks and gladly acknowledge their help. I am especially grateful to the late Mr R. Stone and to Mr E. Pollard who both spent time and trouble reading through the manuscript, helping me to avoid several errors and making sympathetic suggestions on points of detail and style. My former colleague Mr M. Riches willingly gave hours of his time helping to prepare some of the photographs, and my wife spent many evenings producing the typescript; I owe a large debt to each of them.

D.B.

Introduction

This is not a physics textbook. It is more a reader for the general, if not unintelligent, student who wishes to gain some insight into the main branches of the subject. The level of explanation is roughly equivalent to first-examination level, occasionally above that, sometimes below. Not all the topics currently examined by the majority of examination boards are represented here and some digressions are allowed to stray beyond the boundaries of examination syllabuses.

Mathematical skill always makes physics easier to handle, but it is not necessary to be more than competent in mathematics to cope with all the algebra in this book. There are companion volumes in the series dealing with several related topics, and the reader is referred to these if he feels he needs a different style of explanation or an alternative viewpoint to clarify an idea.

The book uses metric (SI) units for all the main sections, though non-SI units do creep in where their size is conveniently right for the occasion, for example volume in cubic centimetres (cm^3). British Imperial units are avoided altogether. The list below gives the basic units and abbreviations used in the book, though quite often the name is written out in full.

Quantity	Unit	Abbreviation
length	metre	m
area	square metre	m^2
volume	cubic metre	m^3
mass	kilogram	kg
time	second, hour	s, h
density	kilogram per cubic metre	$kg\,m^{-3}$
velocity	metre per second	$m\,s^{-1}$

Quantity	Unit	Abbreviation
acceleration	metre per second per second	$m\,s^{-2}$
force	newton	N
pressure	newton per square metre	$N\,m^{-2}$
current	ampere	A
potential difference	volt	V
resistance	ohm	Ω
energy or work	joule	J
power	watt	W
charge	coulomb	C
capacitance	farad	F
temperature	degree celsius, kelvin	°C, K
frequency	hertz	Hz

Multiples and sub-multiples of some of these units are often needed and those most commonly encountered are listed below.

Prefix	Abbreviation	Value		Example
mega-	M	million	10^6	$1\,M\Omega = 1\,000\,000$ ohms
kilo-	k	thousand	10^3	$1\,kW = 1000$ watts
centi-	c	hundredth	10^{-2}	$1\,cm = \frac{1}{100}$ metre
milli-	m	thousandth	10^{-3}	$1\,mA = \frac{1}{1000}$ amp
micro-	μ	millionth	10^{-6}	$1\,\mu C = \frac{1}{1\,000\,000}$ coulomb
nano-	n	thousand millionth	10^{-9}	$1\,ns = \frac{1}{1\,000\,000\,000}$ second
pico-	p	million millionth	10^{-12}	$1\,pF = \frac{1}{1\,000\,000\,000\,000}$ farad

1

Molecules on the Move

1.1 Building bricks

When scientists try to explain about the world around them they often set up ideas of what things *might* be like and then test them by experiment to see if the ideas are any good. Over the years one of the ideas which has worked well is that of **atoms** and **molecules**. We imagine all matter – this book, you and me, the stars, the air, in fact everything – to be made up of millions and millions of tiny building bricks which have been called atoms and molecules.

Atoms are the building bricks of substances which contain only *one* kind of material – lead, oxygen, mercury, zinc, aluminium. One atom of lead is the smallest bit of lead which is still lead; if you break the atom down you do get smaller pieces, but they do not behave like lead any more. Quite often though, several atoms go round together, perhaps similar ones or more commonly groups of different ones, and these groups are called **molecules**. Carbon dioxide has a molecule made up of two atoms of oxygen and one atom of carbon bonded together as a unit forming the smallest possible amount of carbon dioxide. Most everyday substances are made of molecules, groups of atoms, and since there are over 90 different kinds of atoms in or around the earth there will be a vast number of possible combinations. This we see reflected in the countless variety of materials familiar to us – wood, steel, polythene, animal tissue, foods, paints, explosives, etc.

1.2 Has anyone seen an atom?

Since atoms are about a ten millionth of a millimetre across (10^{-7} mm) the most powerful of microscopes will be needed to see them. Fig. 1.1 shows a photomicrograph of a regular array of iron sulphide molecules. A special kind of microscope giving about 20 million times magnification was used to obtain this picture. Even with this magnification the molecules look like so many dots. No detail of them can be resolved, but there does seem to be a regular arrangement of something spaced about 10^{-7} mm apart.

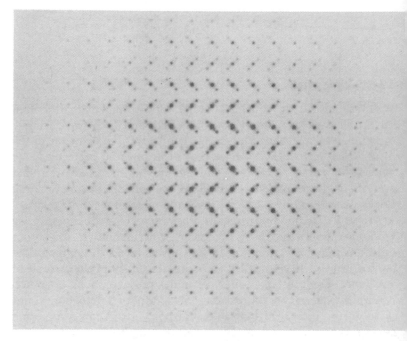

Fig. 1.1 Arrangement of atoms in pyrite, FeS_2, seen looking along an edge of the unit cell. Large dark spots are iron atoms, smaller ones are sulphur atoms. Magnification, about 20 million diameters.

1.3 A simple way to measure molecular sizes

A rough idea of what sizes we are dealing with when we talk of molecules can be obtained using quite simple equipment. You need some olive oil, talcum powder, a clean washing-up bowl, a piece of fine wire, a magnifying glass, a ruler and a piece of card. Fig. 1.2 shows the experiment in stages. Loop the wire and stick it to a small piece of card.

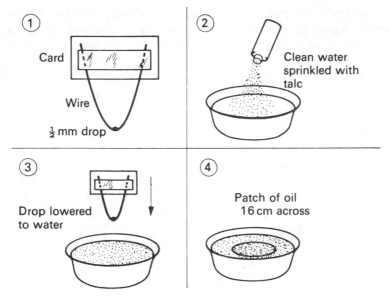

Fig. 1.2

Dip the loop into some olive oil causing droplets of oil to cling to the wire. With another piece of wire scrape all the drops down to the bottom of the loop and then carefully poke the drop until it is $\frac{1}{2}$ mm across. You need the ruler and magnifying glass to judge this.

A $\frac{1}{2}$ mm drop has a volume of nearly $\frac{1}{8}$ mm^3. If this drop is placed on the surface of water it will spread out into a very thin film and some talcum powder, lightly sprinkled on the surface beforehand, will be pushed aside leaving a circular patch about 16 cm across, that is an area of about 200 cm^2.

If the oil layer has a thickness of x mm its volume will be $20\,000x$ mm^3 which must, of course, be equal to the original volume of the drop. This gives

$$20\,000x = \tfrac{1}{8}$$
$$\text{or} \qquad x = \tfrac{1}{160\,000} \text{ mm}$$
$$= 6 \cdot 2 \times 10^{-6} \text{ mm}$$

The olive oil clearly cannot be *thinner* than one molecule thick, so a single molecule will be about this length or smaller. In fact the olive oil molecule is made up of a chain of atoms, so we are still at least 10 or 20 times too long for an estimate of an *atomic* size. There is obviously no chance of either atoms or molecules ever being visible to the naked eye!

1.4 Why are molecules important?

Really it was the chemist who originally concerned himself with the molecular idea because it is the molecules of different substances which interact in chemical reactions. The reasons for substances combining chemically lie with the behaviour of their molecules when close together. Chemical theory became a molecular theory because, by investigating the way these small objects behaved, men were able to understand why bulk matter behaved the way it did and new chemical reactions were suggested. In short, the molecular idea fitted the facts and gave a systematic explanation of what really happened.

That's all very well from a chemist's point of view, but why is it that physicists are equally bothered about molecules? It turns out that many physical properties of substances like boiling point, vapour pressure, hardness, rate of diffusion and several others can be elegantly described in the same molecular terms. This in fact is the only real test of a scientific idea – does it agree with what we observe? – and the molecular theory *does* fit in well. Scientists now accept the notion that all matter is made up of *particles* and is not a smooth mass with no texture or graininess.

We must now consider what we know about the behaviour of these particles (the molecules of matter or, in some cases, parts of molecules) from the point of view of physics.

1.5 Solids, liquids and gases – particle arrangement

Every material we know on earth is either in the form of a solid, or a liquid, or a gas, and a lot of them can be changed from solid to liquid or from liquid to gas without much trouble. Ice becomes water, water turns into steam, iron can be melted, gases can be liquefied and a few will go straight from solid to gas, like carbon dioxide. These changes are generally brought about either by heating or cooling, the three different states having different molecular arrangements.

Solids
Here the material usually has a definite shape and occupies a definite volume. It can often be squashed more or less easily, or stretched, but it

does not need a vessel to contain it. What is it that keeps solid things together? If they are made up of particles of some sort there must be a way of preventing their falling apart – the particles must attract each other to keep the solid's shape. This would also be the reason for it being difficult to tear some solids apart; iron particles must have a bigger attractive force for each other than rubber or paper particles to account for the different strengths of different materials.

So far so good, but it is also very difficult to squash certain solids and if the particles were attracting each other it ought to make it easy to force them together. We have to imagine a situation where if you try to squeeze them closer together the particles react by pushing away from each other, but if you try to pull them apart they change their minds and *pull* one another! An example of that kind of thing would be a set of balls linked to each neighbouring ball by a spring which could be stretched or squashed. The balls will end up a certain distance apart and will resist both being pushed together and pulled apart. Fig. 1.3 is a sketch of such an arrangement.

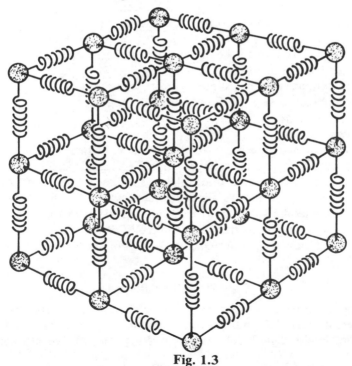

Fig. 1.3

Of course there are no springs between real molecules, but the forces between them act in a similar way to the springs. The molecules settle down to their comfortable distance apart – called their equilibrium positions – and the solid has its definite shape and volume.

Liquids

The situation with liquids is more complex because they have a definite volume but do not mind what shape of vessel contains them. There must still be attractive forces between the molecules or else the liquid would not stay together; and like solids, most liquids are almost incompressible, so the molecules cannot easily be forced together. The attractive forces are not strong enough though to maintain a definite shape.

Generally speaking, liquids occupy a larger volume than the solids they come from; they are less *dense*, so the molecules will on the whole be slightly further apart than in the solid. (The prime exception to this of course is water, which expands when it freezes, but water has a complex molecular pattern and is a freak substance in this respect.) Because the molecules are further apart we would expect the forces of attraction to be somewhat weaker, just as the pull of a magnet decreases as you go further away from it. There must be a rather delicate balance of forces in a liquid – a slightly stronger attraction would produce a solid, a slightly greater separation would mean too little strength to keep the liquid molecules together at all.

Gases

Now we come to the other extreme. Gases have neither shape nor size. They will fill any container they are in and if released will mix with the air so thoroughly that it is impossible to sort the molecules out again. Clearly, taking our picture about molecules further, we must say the gas molecules are so far apart that their mutual attractions become small enough not to matter. Each molecule becomes an independent agent and may be anywhere in the vessel containing the gas.

1.6 Solids, liquids and gases – particle motion

So far the three states of matter have been distinguished mainly by the different molecular spacings – gas much further apart than solid or liquid, liquid slightly further apart than solid. (There are some exceptions to this general statement, but the discrepancies can always be explained in terms of molecular behaviour and are so involved that we will not deal with them here.) There is a much more important difference though between solids, liquids and gases – the molecules are *moving*. The evidence for this will be presented later, but just now let us look at the three states again in terms of the *motion* of the molecules.

Solids

For solids we think that the molecules are arranged in more or less fixed places but yet possess the ability to vibrate. If you can imagine a cube of table tennis balls spaced out by springs as in Fig. 1.3, each ball joined to its neighbours and the whole thing vibrating rather like a jelly, you will have some idea of our view of solids today. The position of each ball is more or less fixed with respect to the other balls, but each ball can vibrate in all directions around its normal position. The general cubic shape is maintained, but to a detailed view the balls are moving about.

This is how we think solids are arranged. The strength of the 'springs' – really the forces between molecules – keeps the material in its general shape, prevents it from being squashed down to almost nothing and acts between neighbouring molecules; yet each molecule has some energy of vibration which causes it to move around its equilibrium position.

When a solid gets hot – assuming no chemical changes, such as burning, take place – more energy is given to the molecules by whatever it is that is heating it. This causes even more vibration and ultimately leads to the breakdown of the regular arrangement of the molecules – a process known as melting, in other words the solid becomes a liquid.

Liquids

This is the most complex of the three states because the molecular motion cannot be so violent that it entirely overcomes the attractive forces between molecules – otherwise the liquid would not stay in a certain volume; but the motion *is* sufficiently agitated for the regular spacing which characterises solid materials not to exist – at least over large volumes or for long periods of time. Probably some local organised sections may arise from time to time in a liquid, but they will soon be broken down by the molecular vibrations only for others to arise elsewhere. This means the molecules are not in fixed places with regard to their neighbours – they may stay near them for a while, but due to collisions with other molecules will wander away from any particular point in a random sort of way, though still under the influence of nearby molecules they meet on their way.

Gases

If still more heat is pumped into the molecules, they gain even more energy and can escape altogether from the attractions of nearby neighbours. The liquid boils. The molecules no longer have merely a vibrating sort of motion, but will be able to move bodily in any direction completely free from any sort of restraint, except when they collide with each other or with the walls of the vessel holding them. Their motion will be quite haphazard, purely random with no direction of travel preferred to any other. This describes a gas in terms of its molecules' motion.

1.7 What evidence is there that molecules actually do move?

Since molecules are very hard to see and certainly no microscope yet developed can hope to show any sort of movement directly, the evidence for molecular motion has to be secondhand. There are two simple consequences of motion to consider – gas pressure and Brownian movement.

Gas pressure

Everyone is familiar with the term 'pressure' applied to the household gas supply – it is the thing which 'pushes' the gas through the pipes from gas holder to cooker. In physics it has a more precise meaning of 'force acting on every unit area', and would be measured in newtons per square metre or millimetres of mercury, but it possesses the everyday notion of 'pushing'. (Chapter 2 deals fully with the concept of force and its units.)

A light ball can be held up by a jet of air from, say, a vacuum cleaner with the pipe attached to the blow end: the airstream supports the ball by pushing against it with the same force as the earth's gravity would bring it down. What keeps it up? The air molecules must bounce off the ball, so it is clear that moving air molecules *can* exert a push.

In a gas that is not moving wholesale, as the output of the blower mentioned above would be, the molecules will be hitting the sides of the vessel holding them from all directions, some striking full square, others glancing off, yet the same mechanism is involved – of molecules bouncing off a surface which feels the impacts as a pressure. In fact we live at the bottom of a 'sea' of air and are being bounced on to the tune of about a hundred thousand newtons per square metre, or enough to hold up a column of mercury 760 mm high. Without the continual movement of the molecules of air there would be no 'atmospheric pressure' like this. A load of 100 kilonewtons every square metre amounts to a fair force over all, but we do not collapse under the load because we have internal pressures which balance the atmosphere and keep us intact. The space travellers and moon walkers who venture into airless regions must surround themselves with a suit strong enough to supply air (or oxygen) at normal pressures – the danger of a leaky suit is that the man inside would literally explode!

Later on when dealing with forces and how they balance (Chapter 2) we shall obtain a fairly simple connection between the pressure exerted by a gas and the characteristics of its molecules – their mass, their total number and their average speed. At this stage it is enough to realise that molecular movement does necessarily, in the case of a gas, cause it to exert a pressure on the walls of the containing vessel and this is the most direct evidence so far for the idea that molecules are moving.

Brownian movement

Although the motion of individual molecules is not observable through a microscope, it ought to be possible to see larger particles being jostled around by the smaller molecules in a mixture or solution. This in fact can be done, but it is not as easy as it sounds. The molecules in an ordinary gas have so little mass that, although they do move very quickly, they do not make a large impact on particles big enough to be seen through a microscope. The 'detector' particles must be small enough to suffer appreciable shifts when hit by the moving molecules, yet large enough for ordinary microscopes to make them easily visible. Nowadays these conditions can be deliberately prearranged without difficulty, but it was originally a lucky and, at the time, puzzling observation. Often major advances in many branches of science have been made quite accidentally while pursuing quite different aims and this happened to Robert Brown in 1827. He was studying pollen grains suspended in a liquid and their ceaseless motion caught his attention.

A simple way to see it nowadays is to trap some smoke from smouldering rope or a cigarette in a small glass tube. Under strong illumination from the side and using a moderately low-powered microscope, the smoke particles look like stars against a dark background and appear to be dancing around all the time for no apparent reason. Each individual particle is being hit many times a second in quite random fashion and its jerky, jittery progress reveals the presence of invisible but *moving* things too small for the microscope to see. (A very high power microscope is no good for this simple arrangement since, although it could easily see the smoke particle, the instrument would have such a small depth of focus that the particles would not be visible long enough for their motion to be observed – they do not of course move obligingly in one plane perpendicular to the microscope tube.) Fig. 1.4 shows the practical arrangement.

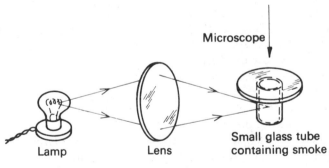

Fig. 1.4

Seen for the first time the effect is startling and the particles may be thought to be alive – a nasty conclusion for a cigarette smoker! – but the movement persists using other definitely inert and passive particles. Increasing the temperature makes the Brownian movement more rapid, agreeing with our tentative ideas of hotter meaning faster-moving molecules, and using larger particles results in a less obvious motion since the larger particles are less easily moved by the bombarding molecules.

1.8 Smaller than atoms?

Before leaving the subject of small particles to deal with large-scale things, the reader will be aware that molecules and atoms can be split into yet smaller bits and these will be mentioned now for fuller treatment later.

As we know already, molecules can be broken down into their constituent atoms, for example a sugar molecule consists of carbon, hydrogen and oxygen atoms. The atoms in their turn can be split into three types of particles – **electrons, protons** and **neutrons** which in normal atoms form a stable arrangement. The electrons and protons carry electric charges, respectively negative and positive,* in equal quantities while the neutron is electrically neutral. The particles differ widely in mass though, a proton and a neutron each being about 1840 times more massive than an electron.

These particles are arranged with the massive ones in the central nucleus and the lighter electrons somewhere outside the nucleus. The scale is quite extraordinary – the electrons are about 100 000 times further from the nucleus than the distance across it! If the nucleus were

1 cm in diameter – this size ◯ – the electrons would be a kilo-

metre away with literally nothing between them. The vast majority of the volume of all materials we call hard or heavy is – empty space!

Just to boggle the reader's mind one step further, it has been suggested in the last few years that these sub-atomic particles may themselves be made up of yet other particles so far unidentified! Big fleas have little fleas . . . little fleas have smaller fleas . . . and so on . . . ? !

1.9 A serious warning

All this section should really be written in the following style – 'matter *behaves as if* it is made of atoms and molecules; these *behave as if* they are made of electrons, protons and neutrons which *behave as if* they could be made of other particles etc. . . . '. The reason for saying this is that these terms are mere descriptions of man-made models, not divine creations, and the reader should try to remember that scientists are concerned to *describe and explain* the world, not to say what it *is*.

* See Chapter 7 for the reasons for these names.

2

Forces and Motion

2.1 Displacement, s

How would you give a friend full instructions for getting from your house, say, to the football ground? You know what you would do as the crow flies, a straight line from here to there. Unfortunately we can rarely travel like that, so you have to give your friend very detailed information such as 'Second on the left after the traffic lights, across the roundabout, fork right by the next telephone box, etc.' The result is that the journey becomes a set of small trips of different lengths and in different directions, but the total effect is the same as if a crow-flight had been possible. In fact all the little trips add up to the single journey going straight to the football ground, and of course it does not matter which of several routes is taken, the total **displacement**, arrow s, is always the same – Fig. 2.1.

The point about this set of directions is that you have to tell your friend not only how far to go but also in which direction. He has to move using a set of **vectors**, the name given to things like this which require a size *and* a direction. Many possible sets of vectors are available, all adding up to the one required. A very convenient way of showing a vector is to use an arrow, the length representing its size and its direction being given by the way the arrow is pointing. In this way we could say arrows s_1, s_2, s_3 add up to give arrow s.

Another way of looking at this is to think of the *change* of displacement between two positions – Fig. 2.2 and Fig. 2.3. Suppose we call the football ground displacement s and the airport displacement r, the displacement between football ground and airport is $r-s$ or

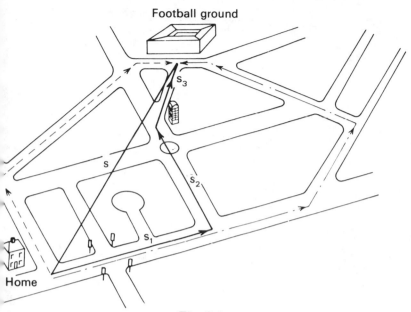

Fig. 2.1

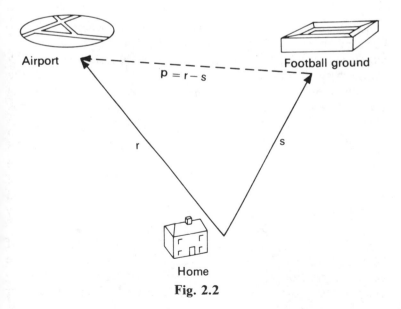

Fig. 2.2

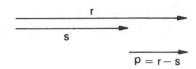

Fig. 2.3

arrow *p*. This clearly works if *p*, *r* and *s* are all in one line, but if we use the arrow idea it can be made to work if they are not in one line. We can get different answers to the same sum if we add up numbers like this – see Fig. 2.4.

$$\overset{\rightarrow}{3} + \overset{\rightarrow}{4} = \overset{\rightarrow}{7}, \text{ or } \overset{\rightarrow}{3} + \overset{\leftarrow}{4} = \overset{\leftarrow}{1}, \text{or } \overset{\rightarrow}{3} + \overset{\uparrow}{4} = \overset{\nearrow}{5}.$$

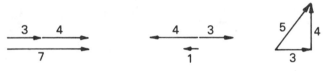

Fig. 2.4

Here is another example of a physical quantity where we need to give some information about direction. To describe the speed of an aeroplane as 600 km h^{-1} does not really help someone who wants to know exactly how the plane has moved, but if you add 'towards the west' the aeroplane's track is known from its starting point. The arrow idea can be used here too. Looking at Fig. 2.5, 800 km h^{-1} westwards added to 600 km h^{-1} northwards gives 1000 km h^{-1} nearly north-westwards. The combination of speed and direction is called the

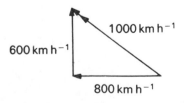

Fig. 2.5

velocity of the aeroplane. It is a measure of how rapidly its position is changing or more precisely it gives its rate of change of displacement. That is why it is quoted in $km\,h^{-1}$ or $m\,s^{-1}$.

The technique of using arrows and adding them up in this way will be needed on several other occasions later in this chapter. In fact any other quantity which requires a size and a direction will have to be treated like this and physics is full of that kind of quantity – velocity, acceleration, force, momentum, etc.

2.2 Velocity and displacement

Suppose every 15 minutes someone reads the speedometer of a car as it travels straight at a steady $60\ km\ h^{-1}$, and then draws a graph of the figures. The result would be like Fig. 2.6. In fact this graph gives information about the car's position or displacement as well as its velocity. How far does the car go in the first 15 minutes? A quarter

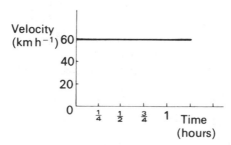

Fig. 2.6

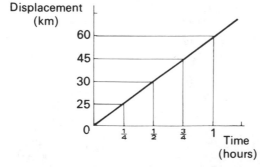

Fig. 2.7

of an hour at $60 \, \text{km h}^{-1}$ means 15 km. How far after 30 minutes? – 30 km. And after 1 hour? – 60 km. Now we can draw the displacement graph too and it looks like Fig. 2.7. The connection between these graphs is that

$$\text{Velocity} = \frac{\text{Displacement}}{\text{Time}},$$

or, Displacement = Velocity × Time.

In symbols:

$$s = ut$$

This result is true as long as the car's velocity remains the same.

2.3 Acceleration, *a*

If the car's velocity were to change it would have to accelerate or decelerate. We need to be able to handle change of velocity or **acceleration**. Suppose a car's velocity is 10, 20, 30, 40, 50 km h^{-1} at quarter-hourly intervals. Its velocity is changing by $10 \, \text{km h}^{-1}$ every 15 minutes, or $10 \, \text{km h}^{-1}$ per $\frac{1}{4}$ hour, or $40 \, \text{km h}^{-1}$ per hour and this is the acceleration number, sometimes written as 40 km per hour per hour or $40 \, \text{km h}^{-2}$. (The commonest example of an accelerating object is one falling freely to the ground; its acceleration then is $9 \cdot 8 \, \text{m s}^{-2}$, i.e. its speed increases by $9 \cdot 8$ metres per second every second.)

The velocity graph of the car now will be like Fig. 2.8, going up

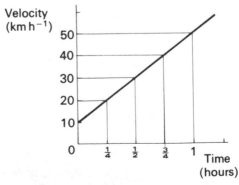

Fig. 2.8

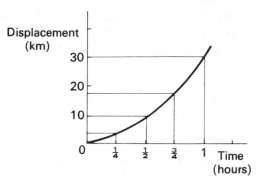

Fig. 2.9

steadily by the same amount for each time interval. What will the displacement graph be this time? The answer is an ever-steepening curve like Fig. 2.9. The car increases its distance from the starting point by more in each equal time interval, as the numbers on the graph show. A mathematical treatment gives us:

$$s = ut + \tfrac{1}{2}at^2$$

If we knew an aeroplane was travelling in a straight line and at a certain time was 100 km from the airport, then 2 hours later it was 500 km away, it would be easy to work out its (average) speed – (500–100) km in 2 hours means 200 km h^{-1}. In a similar way we can work out a car's acceleration if we know two velocities for it and the time taken to change from one to the other. Say a velocity u m s^{-1} at a certain time and a different velocity v m s^{-1} a time t seconds later, then the change of velocity would be $(v - u)$ m s^{-1} in t seconds, or an acceleration of $\dfrac{v - u}{t}$ m s^{-2},

i.e. its acceleration, $\qquad a = \dfrac{v - u}{t}$

or as we usually write: $\qquad \boxed{v = u + at}$

2.4 Momentum

Yet another 'arrow-adding' quantity used in physics is what we call **momentum**. It is a word given a special meaning in physics not really the same as its everyday usage (where it is generally the same as what the scientist calls **inertia**). Physicists mean the result of multiplying mass × velocity; in other words momentum = mv. (We have not really said what **mass** means yet, but we shall deal with that in Section 2.6.) A 75 kg sprinter running at 10 m s^{-1} has a momentum of 750 kg m s^{-1}. Momentum can have a direction just like velocity or displacement and sometimes when dealing with, for instance, the collision of atomic particles or even billiard balls, directions have to be taken into account.

$$\boxed{\textbf{Momentum = Mass} \times \textbf{Velocity}}$$

2.5 Force

The related concepts of force and inertia are two of the most fundamental in physics and until recently required a real leap of faith to accept. The events connected with space exploration have helped greatly to bring these ideas into the everyday experience (at least second-hand experience) of most people.

Imagine a stationary space ship, far out in space. If **no** forces act on the ship what will happen to it? No problem here – it will stay where it is. What if the ship is going at 10 000 km h^{-1} and no forces act on it? Well, it will keep on going in a straight line at 10 000 km h^{-1} since there would be nothing to alter its speed or direction – Fig. 2.10.

Fig. 2.10

Space ships are all very well in imagination, but what about earthly objects – can they demonstrate the same principle? If no force acts, things stay at rest – that is easily seen in ordinary objects. If no force acts, moving things keep moving – this is harder to demonstrate because it is difficult to have movement on earth without at least some frictional forces between the moving parts. It can very nearly be realised by using a small hovercraft such as the 'boat' on a linear air track.

This is a long tube, carefully levelled and closed at one end, with small holes drilled along its length every cm or so, and air blown through from the other end. A light plastic or wooden boat can be supported by the air jets and has a near-frictionless ride along the track if given a gentle push. Fig. 2.11 shows the layout. A white stick is attached to the boat so that it can be photographed by multi-exposure against a black background. The photograph shows the boat in successive positions as it travels down the track and Fig. 2.12 is the result of one experiment where the exposures were $\frac{1}{30}$ second apart.

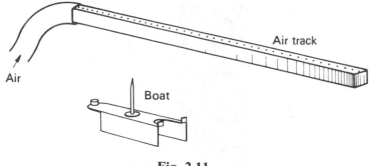

Fig. 2.11

The stick positions are evenly spaced, showing that the boat is in fact travelling at constant speed under the action of no sideways forces. (The other forces – weight of the boat and the upthrust due to the air jets – do not act sideways and are equal and opposite anyway.)

Now without really trying we have said what force is; or at least what a force does. A **force** is something which causes a change in the speed or direction of travel of an object, in other words, it causes an acceleration or deceleration. The force of the rocket engines can drive the space ship forward; the force of attraction of a nearby moon or planet can deflect the ship towards it. Conversely we can say that if a change of velocity is observed there must be a force somewhere responsible for it. No force acting – no acceleration; a force acting – an acceleration (change of speed, or direction, or both).

2.6 Inertia

Every rugby football player knows that a good big 'un will always beat a good little 'un – Fig. 2.13. In a body-contact game the size of the opposition is always something to look out for. A burly forward takes more stopping than a light half back – he has more **inertia**. A double deck bus *can* be pushed by a group of men, but it is much easier to

J. Jardine

Fig. 2.12   Multi-flash photograph of an air-track boat travelling at steady speed

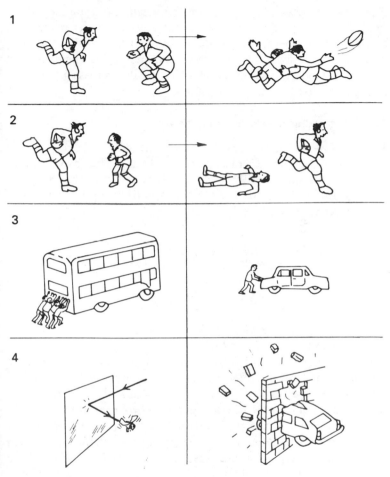

Fig. 2.13

push a mini-car. The bus has more **inertia** than the mini. A collision between a fly and a glass window does little harm to the fly, but a car hitting a brick wall is left a sorry sight. The car has more **inertia** than the fly (also more momentum, which comes into this problem too).

Inertia, then, is the property of 'unshiftability' or resistance to change of velocity. Big objects possess more of it than little ones; but what do we mean by 'big' – size as regards volume or weight or what?

The answer is **mass**. In simple terms this means the amount of 'stuff' in the object, or quantity of material, but it also has a precise meaning as a measure of inertia. A 5 kg lump has half the inertia of a 10 kg lump. It is the object's mass which gives it inertia, and scientists have in Paris a standard lump of metal which is called a mass of 1 kg (see also pages 176–7).

2.7 Force and acceleration – the Newton

How are we going to measure forces? It must be done by the acceleration they produce in a certain sized lump. The obvious size to choose is 1 kg and we will simply say that if a force pulls on a 1 kg lump and accelerates it by 1 m s^{-2} that will be a one-sized force. We call it 1 **Newton** – Fig. 2.14.

1 Newton
Acceleration = 1 ms^{-2}

Fig. 2.14

A force of 1 N speeds up a 1 kg lump by 1 m s^{-2}, but does a force of 2 N speed up a 1 kg lump by 2 m s^{-2}? This is a matter for experiment and measurement – in fact what we want to find is whether doubling the force doubles the acceleration. To do this we need a set of equal forces and some way of measuring acceleration for a truck, say, dragged along a runway. The forces can be provided by equal elastic threads stretched to the same amount and a speedometer read at regular close intervals can give a velocity–time graph like Fig. 2.8. An alternative technique for acceleration measurements is to take a multi-flash photograph of the truck, the exposures being separated by, say, $\frac{1}{25}$th second but all photographed on one frame – Fig. 2.15.

When this experiment is done carefully (and this means making some allowance for friction by tilting the runway slightly or using some frictionless method such as a linear air track or dry-ice puck arrangement) the results look like Fig. 2.16. Doubling the force *does* double the acceleration and this relationship between quantities is called *direct proportion* (sign $\propto$).

Acceleration $\propto$ Force

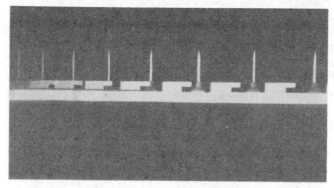

M. Riches

Fig. 2.15 Multi-flash photograph of an accelerating air-track boat

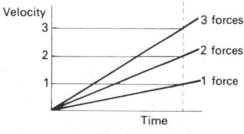

Fig. 2.16

What happens if instead of pulling the same truck with different forces we pull different trucks with the same force? Another experiment will settle this, similar to the last one but using only one elastic and a series of different-sized trucks. The easiest thing to do is to use

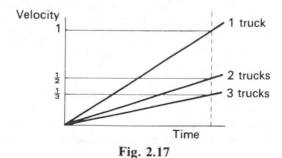

Fig. 2.17

trucks all the same size and stack them on top of each other to make bigger trucks. If the trucks are all of equal mass we will simply be doubling or trebling the mass without needing to know what the mass actually is.

This time the result looks like Fig. 2.17. Doubling the mass halves the acceleration, trebling the mass cuts it down to $\frac{1}{3}$, etc. We say mass and acceleration are *inversely proportional* to each other, or the acceleration varies inversely with the mass.

$$\text{Acceleration} \propto \frac{1}{\text{Mass}}$$

2.8 The full equation

If we use a bit of mathematical help here we can combine the two relations above into one:

$$\text{Acceleration} \propto \frac{\text{Force}}{\text{Mass}}$$

or: Force $\propto$ Mass $\times$ Acceleration.

Going one step further we can make this into an equation:

Force = Mass $\times$ Acceleration $\times$ a number.

The number very conveniently becomes 1 **if** we use newtons to measure the force, since we have already said that 1 newton gives 1 acceleration to 1 mass. So the simple result is:

$$F_{\text{(newtons)}} = m_{\text{(kg)}} \times a_{\text{(m s}^{-2})}$$

We have already decided (Section 2.4) what happens when an object is *not* acted upon by a force. Now we know what happens when there is a force acting – it causes the object to accelerate according to this equation and, of course, the direction of a is the same as the direction of F.

2.9 The earth's pull

The earth pulls all objects towards it with the same acceleration of $9.8 \, \text{m s}^{-2}$. Thus on a 1 kg mass the force must be 9.8 newtons. On a

2 kg mass it is 2×9.8 newtons or 19.6 newtons. On a 75 kg mass the force is 75×9.8 newtons or 735 newtons. Or we can say simply that the earth has a gravitational 'strength' of 9.8 newtons per kilogram. The moon's 'strength' is only about 1.6 newtons per kilogram (see also pages 176–7).

2.10 Forces go in pairs

What we have discussed in the last few sections is relevant if there is a net or resultant or unbalanced force acting on an object – the thing accelerates in the direction of the force. Yet forces never occur singly. Why should the earth pull you downwards when you fall and you not pull the earth upwards? If every piece of material attracts every other piece this should happen – and it does! But of course the movement of the earth upwards is so slight that we cannot normally measure it – Fig. 2.18.

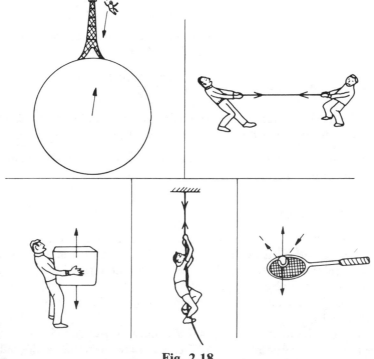

Fig. 2.18

When you have a tug of war with a friend you cannot pull him without his also pulling you. You cannot lift up a box of bricks without their pushing downwards on your arms. You cannot climb up a rope without pulling the rope downwards. A ball being hit by a tennis racket also gives the racket a jerk. A bird can be hit by an aeroplane and be killed, but it will make quite a dent in the plane.

All these are examples of forces occurring in pairs and you can easily think of many others. Why don't you fall through the floor? How can a hovercraft hover? How do space ships alter course in space?

This is the third main discovery about forces: whenever a force acts, there will also be an equal and oppositely directed force acting elsewhere.

2.11 Summary

I If no force acts, or if all the forces are balanced, an object will stay still or carry on going at constant speed in a straight line.

II If an unbalanced force acts, an object will accelerate in the direction of the force according to the relation

$$\text{Acceleration (in m s}^{-2}) = \frac{\text{Force (in newtons)}}{\text{Mass (in kg)}}$$

III Whenever a force acts on an object, an equal and opposite force acts on the agent producing the original force.

These three statements are known as **Newton's Laws of Motion** and are the basis of all mechanics.

2.12 Bomb dropping

A convincing demonstration of Laws I and II both working at the same time is the path of a bomb or bullet. Think of an aeroplane cruising along at steady speed in a straight line at, say, 140 m s^{-1}. Any bomb it carries also has that horizontal speed and as long as no horizontal forces act on it the bomb will retain the same 140 m s^{-1} horizontally.

When the bomb is released it will also have a vertical acceleration of 9.8 m s^{-2}, but the point is that the vertical force due to the earth's pull only produces a vertical acceleration, it does not affect the horizontal 140 m s^{-1} at all. In equal intervals of time the bomb goes equal

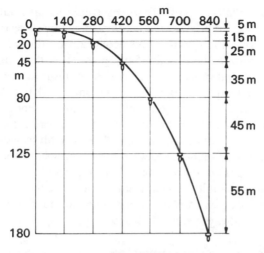

Fig. 2.19

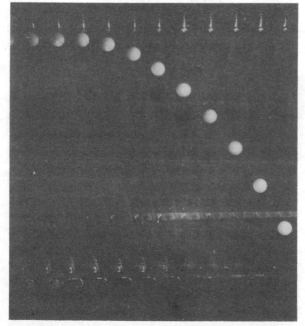

J. Jardine

Fig. 2.20 Multi-flash photograph of a falling ball

distances horizontally, 140 m every second, but vertically the distances increase every second because it is accelerating in this direction. In the first second it goes about 5 m, in the second second about 15 m, in the third about 25 m, in the fourth 35 m, in the fifth 45 m, in the sixth 55 m, etc. The graph of Fig. 2.19 shows these positions every second and the dotted line is the curve taken by the bomb as it drops. One interesting thing is that it is always underneath the aeroplane – the pilot will have to veer off if he wants to avoid the explosion, especially on low-level flights! (This analysis is not strictly true in practice because there is always some air resistance which slows down both the horizontal and vertical motions.)

Fig. 2.20 shows a multi-exposure photograph of a ball released from a carriage travelling at constant speed. The ball traces out the curve expected and is always vertically below the carriage.

2.13 A fable – on safari for a monkey

One day a hunter went out hunting monkeys. Soon he found one hanging quietly from the branch of a tree. The hunter was not good at physics and he argued that if he aimed the gun directly at the monkey he would be sure to hit it.

Fig. 2.21

But the monkey had also seen the hunter and as the hunter raised his gun the monkey thought carefully what to do. He decided he would let go of the branch just as he saw the hunter's finger squeeze the trigger, and then he thought the bullet would miss him.

So ... bang ... jump ... one dead monkey! The hunter was astonished but pleased, and realised the mistake he had made. The monkey was dead and could not realise the mistake *he* had made. (Both the monkey and the bullet accelerate downwards at exactly the same rate, of course! – Fig. 2.21.)

2.14 Force and momentum

Here is a short piece of algebra using the symbols for force, mass, acceleration, velocity and time from Sections 2.3 and 2.9.

$$a = \frac{v - u}{t} \text{ and } a = \frac{F}{m}.$$

Combining these gives

$$\frac{F}{m} = \frac{v - u}{t}$$

and rearranging,

$$F = \frac{mv - mu}{t}.$$

In Section 2.4 the product mass × velocity was called momentum, so we can write

$$\boxed{\text{Force} = \frac{\textbf{Change in momentum}}{\textbf{Time taken for the change}}}$$

Example 1

An 80 kg man travelling at 1 m s^{-1} hits a foam rubber mattress and is stopped in 2 seconds. What force does he feel?

$$\text{Force} = \frac{\text{Change in momentun}}{\text{Time taken}}$$

$$= \frac{80 \times 1 - 80 \times 0}{2} = \underline{40 \text{ newtons}}.$$

Example 2

An 80 kg man walking at 1 m s^{-1} collides with a brick wall and is stopped in $\frac{1}{20}$ second. What force does he feel?

$$\text{Force} = \frac{\text{Change in momentum}}{\text{Time taken}}$$

$$= \frac{80 \times 1 - 80 \times 0}{\frac{1}{20}} = \underline{1600 \text{ newtons!}}$$

These examples point the moral of wearing seat belts in cars, which allow any rapid deceleration of a human body to be taken over a longer time rather than a short and very sharp collision with the dashboard!

2.15 What happens to momentum in collisions?

This important section will be worked through in numbers and also in symbols, which appear alongside the numbers they could represent.

Imagine a billiard ball, mass 0·2 kg travelling at 4 m s^{-1}, following a bigger ball of mass 0·3 kg travelling in the same direction at 2 m s^{-1}. It catches up with the bigger ball and after colliding the bigger ball is still moving in the same direction but at 3 m s^{-1}. What happens to the smaller ball? Suppose the time of contact is $\frac{1}{10}$ second. During this time each ball will act on the other with a force F (two equal and opposite forces, Newton Law III). Fig. 2.22 describes the problem.

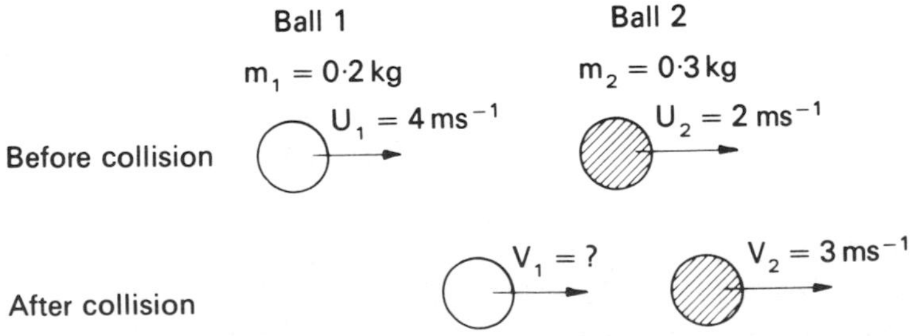

Ball 1   Ball 2
$m_1 = 0\!\cdot\!2$ kg $m_2 = 0\!\cdot\!3$ kg

Before collision $U_1 = 4$ ms^{-1} $U_2 = 2$ ms^{-1}

After collision $V_1 = ?$ $V_2 = 3$ ms^{-1}

Fig. 2.22

Change in momentum of
   ball 2 $= 0.3 \times 3 - 0.3 \times 2 \text{ kg m s}^{-1}$ $m_2 v_2 - m_2 u_2$
       $= 0.3 \text{ kg m s}^{-1}$

Time taken $= \frac{1}{10}$ second
Force acting,

$F \quad = \dfrac{0.3}{\frac{1}{10}}$ newton $F = \dfrac{m_2 v_2 - m_2 u_2}{t}$

   $= 3$ newtons.

Force acting on ball 1 is also 3 newtons but in the opposite direction.

Acceleration of ball 1
   in reverse direction $= \dfrac{3}{0.2} \text{ m s}^{-2}$ $a = -\dfrac{m_2 v_2 - m_2 u_2}{m_1 t}$

   $= 15 \text{ m s}^{-2}$

Change of velocity
   in $\frac{1}{10}$ second $= 15 \times \frac{1}{10} \text{ m s}^{-1}$ $v_1 - u_1 = at$
   $= 1.5 \text{ m s}^{-1}$
New velocity of ball 1 $= 4 - 1.5 \text{ m s}^{-1}$ $v_1 = u_1 - \dfrac{m_2 v_2 - m_2 u_2}{m_1}$

   $= \underline{2.5 \text{ m s}^{-1}.}$

How much momentum did the balls have before and after their collision?

Before: $(0.3 \times 2 + 0.2 \times 4) \text{ kg m s}^{-1} = 1.4 \text{ kg m s}^{-1}$
After: $(0.3 \times 3 + 0.2 \times 2.5) \text{ kg m s}^{-1} = 1.4 \text{ kg m s}^{-1}$

so no momentum was lost or gained, it was merely redistributed.
   If we rearrange the algebraic equation on the right above we get:

$$\boxed{m_1 u_1 + m_2 u_2 = m_1 v_1 + m_2 v_2}$$

which says the same thing in symbols.

2.16 Conservation of momentum

The usual way in physics to indicate that a quantity does not alter
when other changes occur is to say it is **conserved**.

In any system of colliding objects the total amount of momentum at any time remains fixed. This statatement is usually called the Conservation of Momentum (strictly speaking *linear* momentum since we have not allowed for spin) and it is thought to apply to all manner of collisions, from billiard balls to molecules to atomic particles to meteors and even to stars. It is one of the most important ideas in physics.

(If you have your wits about you, you will realise that the truth of it depends on the two forces of ball 1 on ball 2 and of ball 2 on ball 1 being equal – i.e. Newton III is an alternative way of saying the same thing.)

2.17 Momentum in real collisions

In Section 2.5 a technique using a linear air track and multi-exposure photographs was explained to enable almost friction-free motion to be obtained and recorded. The same method can be used to show that the result of Section 2.15 about momentum in collisions does in fact work out in real life.

Two boats are needed this time, one carrying an elastic band and the other a projecting bit which will hit the elastic band when the boats collide – Fig. 2.11 shows a possible design for a boat. If the boats are of unequal mass we can investigate the total momentum before and after collision by photography as before. Fig. 2.23 indicates what happens to start with and the camera is arranged so that it photographs the top half of the sticks carried by the boats as they approach before collision, and the bottom half of the sticks as they separate after collision. One such photograph is shown in Fig. 2.24, for a boat of mass 3 units on the left and one of mass 2 units on the right.

Taking 1 cm on the photograph to represent a velocity of 1 m s^{-1} the reader can work out the figures for the total momentum before and after collision, remembering to count momentum in one direction positive and in the other direction negative.

Fig. 2.23

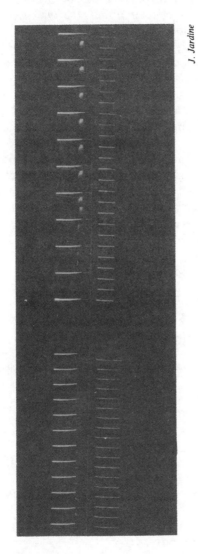

J. Jardine

Fig. 2.24 Multi-flash photograph of colliding boats on a linear air track

2.18 An important calculation for gases

The connection between force and momentum enables us to test our ideas of the previous chapter about the way gas molecules are continually on the move. This motion must be able to cause the observed atmospheric pressure under which we live. The steps in the argument will be set out side by side with the symbols on the right and numbers on the left.

Imagine a box measuring 3 m by 4 m by 6 m, as in Fig. 2.25, containing 3000 small ball bearings each of mass $\frac{1}{2}$ g or $\frac{1}{2000}$ kg. Suppose the ball bearings are bouncing around randomly inside the box, hitting each other and the sides of the box like perfect rubber balls, losing no speed in collisions. Let their speed, whichever way they are going, be 12 m s^{-1}. Imagine *one* ball bearing approaching the end face *PQRS* at this speed and hitting it full on. It will bounce off again at exactly the same speed, travel to the opposite end face *WXYZ* and return after travelling the 12 m in just 1 second.

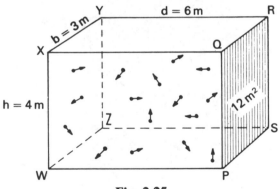

Fig. 2.25

Momentum of the ball bearing before hitting face *PQRS*	$= \frac{1}{2000} \times 12 \ \text{kg m s}^{-1}$	mv
Momentum after hitting face *PQRS*	$= -\frac{1}{2000} \times 12 \ \text{kg m s}^{-1}$	$-mv$
Change in momentum on hitting face *PQRS*	$= 2 \times \frac{1}{2000} \times 12 \ \text{kg m s}^{-1}$	$2mv$
	$= \frac{12}{1000} \ \text{kg m s}^{-1}.$	

This change in momentum occurs every second, so the force it causes will be given by

$$\text{force} = \tfrac{12}{1000} \text{ newtons.} \qquad\Big|\qquad F = \frac{2mv}{t}$$

This will not be the only ball bearing to hit face *PQRS* in 1 second. How many of the 3000 will do so? If they are moving as many in one direction as in any other, a simple way out would be to assume that at any instant $\tfrac{1}{6}$ of them are travelling perpendicular to each face of the box. Thus $\tfrac{1}{3}$ of the total will be involved in collisions with *PQRS*, the $\tfrac{1}{6}$ moving towards *PQRS* and the $\tfrac{1}{6}$ moving towards *WXYZ*.

Total force on face
PQRS due to $\tfrac{1}{3}$ of
the balls $\qquad = \tfrac{1}{3} \times 3000 \times \tfrac{12}{1000}$ newtons. $\qquad\Big|\qquad F = \tfrac{1}{3}n\dfrac{2mv}{t}$

The pressure on the face *PQRS* is the force acting per square metre,

$$\text{pressure} = \frac{12}{4 \times 3} \text{ newtons per} \qquad\Big|\qquad p = \frac{2nmv}{3tbh}$$
$$\text{square metre}$$

$$= \underline{1 \text{ N m}^{-2}}$$

If we had worked this out for any of the other faces we would have reached the same figure for the pressure. The algebraic formula can be made simpler by writing the time between collisions, $t = \dfrac{2d}{v}$. Putting this in gives:

$$\text{pressure} \qquad p = \frac{2nmv^2}{6dbh}$$

$$\text{or:} \qquad \boxed{p = \tfrac{1}{3}\frac{nmv^2}{V}}$$

where V is the total volume *dbh* of the box.

(Just to check, try the figures in the formula for p to see if we get the same answer:

$$p = \tfrac{1}{3} \times 3000 \times \tfrac{1}{2000} \times \frac{12 \times 12}{3 \times 4 \times 6} \text{ N m}^{-2}$$

$$= 1 \text{ N m}^{-2})$$

In the case of a gas, of course, there are many more than 3000 molecules even in a thimbleful and they are not obligingly going in three perpendicular directions nor necessarily contained in a rectangular box, but if the working out is done properly the result is the same as that above except the symbol v stands for an *average* velocity of the molecules.

Let us see what this means for a gas. Normal air exerts a pressure at sea level equal to that of a column of mercury about 0·75 m high. A chunk of mercury 1 m by 1 m by 0·75 m has a mass of 10 200 kg and so the earth pulls it downwards with a force of $10\,200 \times 9\cdot8$ newtons or 100 000 newtons. This gives a pressure on the 1 square metre base of 100 000 newtons per square metre, so we know the size of the pressure, p. What about the product $\dfrac{nm}{v}$? This is just the total mass of air, nm, in a certain volume V and direct weighing gives this to be about 1·2 kg per cubic metre.

We can put these values into the formula for p and work out the average velocity of the molecules, v.

$$100\,000 = \tfrac{1}{3} \times 1\cdot2 \times v^2$$

so

$$v^2 = \frac{300\,000}{1\cdot2} = \frac{1\,000\,000}{4}$$

which gives

$$v = \frac{1000}{2} \text{ metres per second}$$

or

$$v = 500 \text{ m s}^{-1}$$

This is a fantastic result! Ordinary air molecules must be jumping around at supersonic speeds if they are to give rise to the measured value of real atmospheric pressure. This figure is in fact quite realistic and is in good agreement with direct methods of measuring the speeds of molecules. No wonder Brownian movement is so easily seen, the particles must suffer quite a jerk at each collision with an air molecule.

2.19 Going round in circles

How is it possible for the earth to carry on orbiting the sun for ever and ever without apparently anything to drive it round? Why does the moon not fall and hit the earth? How do artificial satellites stay up?

Why must cyclists lean into a corner whereas a car seems to want to go outwards? How do spin-dryers spin dry? All these are examples of circular motion and the special properties of this kind of movement must be examined now.

We will concern ourselves with a constant rate of going round – a steady *angular velocity* or a fixed number of revolutions per second. You might think that this condition is covered by Newton's Law I (Section 2.11), but in fact that situation is one of constant velocity in a *straight* line, not in a circle. If you go round in a circle you might well have a constant *speed*, but certainly not a constant velocity since this involves direction as well as size and the direction is continuously changing.

Think of a stone on the end of a string whirled round in a circle by a boy – Fig. 2.26. At a certain time it will be position *A* of Fig. 2.27 and going at that instant with a velocity *v*, say, in the direction *AX*. A little later on the stone is at *B* with a velocity still of size *v* but this time in

Fig. 2.26

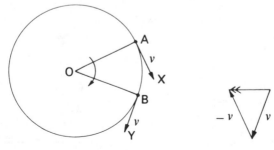

Fig. 2.27

direction BY. By Newton's Law II (Section 2.11) we know that if the velocity alters at all there must be an unbalanced force causing the change; in other words the stone is being accelerated even though it is going round in a circle at constant speed! Which way is it being accelerated?

Imagine first of all two velocity vectors AX 3 m s^{-1} and BY 4 m s^{-1} in the same direction. Suppose the velocity changes from vector AX to vector BY in 1 second; the difference between them gives the change of velocity in 1 second. If AX and BY are in line we get a change of 1 m s^{-1} or 7 m s^{-1} in 1 second depending on the directions of AX and BY. If

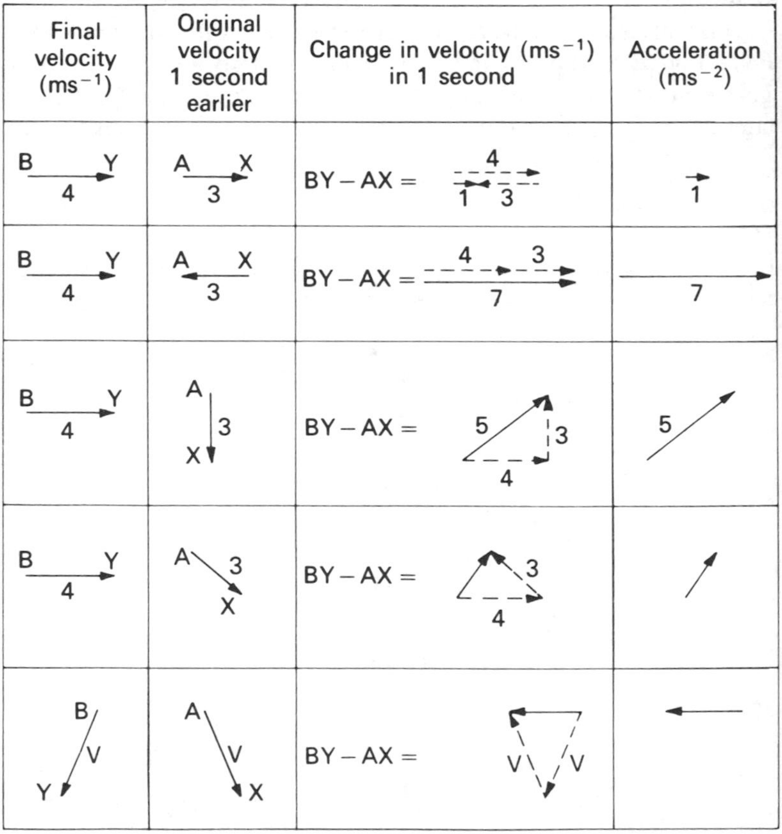

Final velocity (ms^{-1})	Original velocity 1 second earlier	Change in velocity (ms^{-1}) in 1 second	Acceleration (ms^{-2})
B ⟶ Y 4	A ⟶ X 3	BY − AX = (4 over 1,3)	→ 1
B ⟶ Y 4	A ⟵ X 3	BY − AX = (4 → 3) / 7	⟶ 7
B ⟶ Y 4	A ↓ X 3	BY − AX = 5, 3, 4	↗ 5
B ⟶ Y 4	A ↘ X 3	BY − AX = 3, 4	↗
B↙ V Y	A↘ V X	BY − AX = V V	⟵

Fig. 2.28

AX and *BY* are at right angles we still have to subtract them to find the change *BY–AX* which in this case is 5 m s^{-1}, as shown in the Fig. 2.28. With *AX* and *BY* not at right angles the size of the difference *BY–AX* is not obvious, but the way of working it out is the same as for the other cases. What we have to do is to add (*–AX*) to *BY*.

For circular motion we have *AX* and *BY* equal in size but in slightly different directions. The arrow showing the change of velocity in 1 second points inwards towards the centre of the circle. In other words the stone on the end of the string is being accelerated inwards, which in turn means there must be a force producing the acceleration towards the centre of the circle. The pull of the string provides this force; without it the stone would fly off along a tangent to the circle.

2.20 Centripetal force

The argument of the last section is true for any object describing a circular path – it *must* have an inwards-directed acceleration and something somewhere must be providing an unbalanced force in that direction, towards the centre of the circle.

The string clearly provides the inwards force in the case of the stone. For the earth going round the sun it is the sun's gravitational pull which obligingly maintains our orbit year after year. The earth's pull likewise keeps the moon and other earth satellites in their orbits. (Another way of looking at it is, say, that the satellites 'fall' towards the earth from their straight line paths just enough to follow a circular path – Fig. 2.29.)

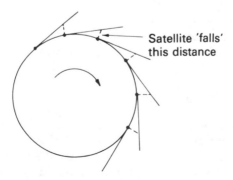

Satellite 'falls'
this distance

Fig. 2.29

Friction between the tyres and the road gives a car or a bicycle their inwards forces. The trouble is that these frictional forces do not act through the centre of the car or bicycle and therefore try to tip them up – Fig. 2.30. The car pivots slightly on its outer wheels and raises its centre of gravity enough for the weight of the car to provide an opposite twisting effect. A cyclist, on the other hand, consciously leans into a curve so that his weight again compensates for the tipping action of the frictional force.

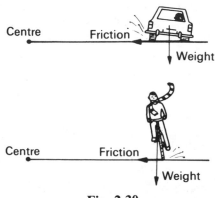

Fig. 2.30

The drum of a spin-dryer gives the inwards force to keep the clothes in it rotating, but there are holes in the drum where there is nothing to provide the force. Water, being free there to follow the normal straight line motion of objects with no force, gets through the holes and leaves the clothes behind in the drum.

The general term for a centre-directed force, which is common to all the examples discussed in this section, is a **centripetal** force. The word means 'seeking the centre'.

2.21 What about centrifugal force?

Many people believe themselves forced *out*wards when going round in a circle – how else does a centrifuge work? Flywheels of old steam engines were sometimes known to disintegrate due to this 'centrifugal' force. Anyone who has been on the Rotor at a fairground will also be likely to believe in this outwards force, the thing which keeps you firmly pinned to the wall while the floor drops away! There is no doubt that in the Rotor you *do* feel yourself being flung outwards, yet according to Section 2.19 all we require for circular motion is an inwards force! How do we resolve this problem?

The answer is that it all depends on where you view things from. Looked at from the outside the forces acting on a Rotor participant are his weight, a large frictional force (equal to his weight) between him and the foam rubber lining of the walls, and an inwards force from the foam rubber which is clearly compressed by the person when he is spinning round. This is the view we normally take of a rotating object; looked at from the outside there is only an *in*wards force, not an *out*wards one.

But from the point of view of the man *in* the Rotor things look different. He feels happily at rest against the wall as long as it is going fast enough for him not to slide down, and remembering Newton's Law I (Section 2.11) the man will consider there to be no unbalanced forces acting on him. Since he feels the wall pushing him in the back all the time he decides there must be another equal force to this to 'balance' it. This invented force he calls centrifugal force and it is real enough to the man as long as he considers himself to be at rest in the rotating room. To the clothes spinning round inside a dryer the water appears to go outwards through the holes in the drum due, the clothes would say, to centrifugal force. The outsider though sees the water going off along tangents to the drum in the absence of any centripetal force to make it go round with the clothes.

2.22 How big is centripetal force?

To find the size of a force we need to know the mass and acceleration of the object on which it acts, Newton's Law II (Section 2.11). Acceleration means rate of change of velocity (Section 2.3) so we have to look at the velocities of a massive object undergoing circular motion on two occasions very close together to find how fast the velocity is changing. Fig. 2.31 shows the two velocities, each v m s^{-1}, represented by AX and BY. The difference BY–AX is represented by the line BX in Fig. 2.32, and if the positions A and B were t seconds apart the acceleration will be $\dfrac{BX}{t}$ m s^{-2}. How long is BX?

If A and B were *very* close together so that AX and BY were almost

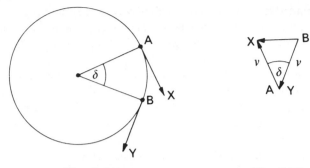

Fig. 2.31   **Fig. 2.32**

parallel (i.e. if t were very small), we could say that the angle δ in radians is given by $\dfrac{BX}{v}$.

$$\delta = \frac{BX}{v}$$

δ is the angle turned through in t seconds. Another way of putting angle δ also in radians is, from Fig. 2.31,

$$\delta = \frac{\text{Arc } AB}{\text{Radius, } R}$$

The arc AB must be of length vt metres since the object has travelled at $v \, \text{m s}^{-1}$ for t seconds. Equating these two expressions for δ gives:

$$\frac{BX}{v} = \frac{\text{Arc } AB}{R} = \frac{vt}{R}$$

or $\qquad BX = \dfrac{v^2 t}{R}$

so the acceleration $\qquad \dfrac{BX}{t} = \dfrac{v^2}{R}$

Consequently the centripetal force, F_{cent}, on a mass m kg going round in a circle of radius R m at a speed $v \, \text{m s}^{-1}$, will be given in newtons by:

$$\boxed{F_{\text{cent}} = \frac{mv^2}{R}}$$

(If the reader finds this proof difficult he is advised not to worry, the result is the important thing and at least he will have been exposed to the method. A friendly mathematician may explain what radians are and how to get over the difficulty of A and B having to be very near to each other. Despite this the result is correct.)

Angular speed is often quoted in 'revs per second' so we will write the result just obtained in terms of this for convenience. One revolution of a circular path means a distance of $2\pi R$ metres, so if there are n revs per second the distance covered in 1 second will be $2\pi Rn$ m. This of course is the speed v. Putting $v = 2\pi Rn$ into the result for F_{cent} gives:

$$F_{cent} = \frac{m(2\pi Rn)^2}{R}$$

or $\boxed{F_{cent} = 4\pi^2 n^2 mR}$

This means that doubling the rate of revolution would quadruple the force required and doubling the radius would double the force.

2.23 Some typical centripetal forces

(a) Take the case of a 1 kg stone being whirled round on the end of a string 1 m long just once every second.

$$F_{cent} = 4\pi^2 \times 1 \times 1 \times 1 \text{ newtons} \approx \underline{40 \text{ N}}$$

(b) For a 500 kg car negotiating a bend of radius 100 m at 13 m s^{-1} (about 47 km h^{-1} or nearly 30 mph) there must be an inwards force of:

$$F_{cent} = \frac{500 \times 13^2}{100} \text{ newtons} \approx \underline{850 \text{ N}}$$

(c) For a gram of water spinning round in a spin dryer, drum radius 40 cm, 4 times every second, the force would be:

$$F_{cent} = 4\pi^2 \times 0.4 \times 16 \times \tfrac{1}{1000} \text{ newtons} \approx \underline{\tfrac{1}{4} \text{N}}$$

The absence of anything to provide this force allows the water to leave the drum.

(d) For an artificial satellite orbiting the earth we can work out its time of revolution if we know its height. The inwards force is provided by the satellite's weight, and if the orbit is a low one, say 150 km high, we can still take the earth's pull to be 9·8 N kg⁻¹ (though in fact it will be a little lower). Let the mass of the satellite be M kg, then using Fig. 2.33,

$$9\cdot8 \times M = \frac{Mv^2}{6550 \times 1000}$$

since the earth's radius is 6400 km.

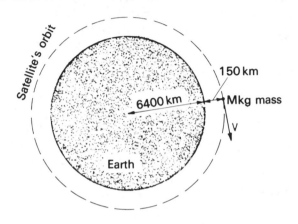

Fig. 2.33

This gives $v = \sqrt{9\cdot8 \times 6550 \times 1000}$ m s⁻¹

≈ 8000 m s⁻¹ or 8 km s⁻¹

or 5 miles per second
or 18 000 mph.

The time taken for one revolution of 2π6550 km is then $\dfrac{2\pi 6550}{8}$

seconds. This is about 4800 seconds or 1 h 20 min.

To compare a real satellite with this result is not as easy as it seems since most artificial satellites travel in elliptical rather than circular orbits. The figure is quite close to the 88·8 minutes it took Gemini 4 to orbit the earth and that satellite's height varied between 162 and 281 km.

(*e*) For the earth's yearly journey around the sun an enormous force is needed. The earth's mass is about 6×10^{24} kg, its orbit has a radius of 150×10^9 m, and it goes round once in a year!

$$F_{cent} = \frac{4\pi^2 \times 6 \times 10^{24} \times 150 \times 10^9}{(365 \times 24 \times 60 \times 60)^2} \text{ newtons}$$

$$\approx \underline{3 \cdot 5 \times 10^{23} \text{ N}}$$

(*f*) One suggested scheme for a manned space station is to design it like a 'doughnut' with a square 'room' running round the hollow tube like Fig. 2.34. The whole device would be made to spin about the central point 0 in order to provide an artificial gravity. Suppose the doughnut were 50 m in diameter, how fast would it have to go round to simulate a normal gravitational field of $9 \cdot 8 \text{ N kg}^{-1}$?

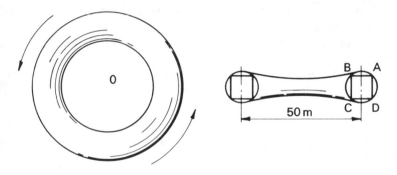

Fig. 2.34

The centripetal acceleration is $\dfrac{v^2}{R}$ or $4\pi^2 n^2 R$ from Section 2.22. We must therefore arrange that

$$9 \cdot 8 = 4\pi^2 n^2 \times 25$$

or

$$n = \sqrt{\frac{9 \cdot 8}{4\pi^2 25}} \text{ rev per second}$$

$$\approx \underline{\tfrac{1}{10} \text{ rev per second.}}$$

This is a surprisingly low result, showing the idea to be a practical one.

Rockets mounted tangentially could initiate the rotation. *AD* would, of course, be the 'floor' of the room and *BC* the 'ceiling'.

2.24 The moon's motion

A few simple observations about the moon's eclipses can lead us to an important result concerning the force of gravity.

The moon's size is very nearly $\frac{1}{2}°$, that is to say the angle from the edges of the moon to the earth is $\frac{1}{2}°$ – Fig. 2.35 – and during an eclipse of the sun the moon's shadow only just reaches the earth's surface. We know this because the patch of total darkness cast by the moon's shadow is only a few miles wide. When an eclipse of the *moon* occurs though, as in Fig. 2.36, the earth's shadow is much wider than the moon and the moon is obscured for some time. By measuring this time it is easily worked out that the earth's shadow is $2\frac{1}{2}$ moon diameters wide where the moon crosses it and that the earth itself is therefore $3\frac{1}{2}$ moon diameters across. This, combined with the $\frac{1}{2}°$ angle, enables us to work out by geometry that the moon is about 60 times the earth's radius away from the earth.

Let us try to repeat the satellite calculation for the moon just as in the last section to find how long the moon should take for one orbit of

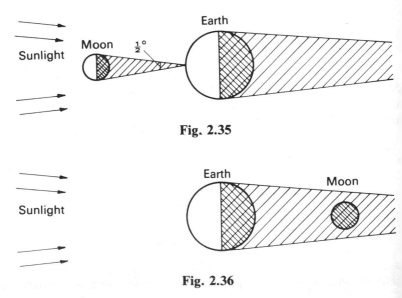

Fig. 2.35

Fig. 2.36

the earth. If the moon's mass is M kg we have

$$9 \cdot 8 \times M = \frac{Mv^2}{60 \times 6400 \times 1000}.$$

The orbital speed $v = \sqrt{60 \times 6400 \times 1000 \times 9 \cdot 8}$ m s^{-1}

The time for one revolution is the distance round ($2\pi R$) divided by the speed v, so the time of orbit,

$$T = \frac{2 \times \pi \times 60 \times 6400 \times 1000}{\sqrt{60 \times 6400 \times 1000 \times 9 \cdot 8}} \text{ seconds}$$

$$\approx \underline{11 \text{ hours!}}$$

Obviously there is something wrong here since the moon really takes about 27·3 days for one orbit. Where is the mistake?

We used $9 \cdot 8$ N kg^{-1} for the strength of the earth's pull on the moon, the value we know it exerts at its surface; but will it still exert $9 \cdot 8$ N kg^{-1} some 60 times further away? Surely not. Perhaps it will be 60 times weaker, or $\frac{9 \cdot 8}{60}$ N kg^{-1}. Trying this gives:

$$T = \frac{2 \times \pi \times 60 \times 6400 \times 1000}{\sqrt{60 \times 6400 \times 1000 \times \dfrac{9 \cdot 8}{60}}} \text{ seconds}$$

$$\approx \underline{3 \cdot 5 \text{ days!}}$$

Still not correct. Evidently the earth's pull must be even weaker than this. Newton suggested that it decreased not by 60 times but by 60×60 times or by a factor of 60^2. Using this in the calculation gives:

$$T = \frac{2 \times \pi \times 60 \times 6400 \times 1000}{\sqrt{60 \times 6400 \times 1000 \times \dfrac{9 \cdot 8}{60 \times 60}}} \text{ seconds}$$

$$\approx \underline{27.4 \text{ days.}}$$

Right this time. This kind of consideration led Newton to 'guess' his universal law of gravitation giving the mutual attractive force F between any two masses M_1 and M_2 when they are separated by a distance R as:

$$F \propto \frac{M_1 M_2}{R^2}$$

The number required to make this an equation is usually written as G, so we write:

$$\boxed{F = G \times \frac{M_1 M_2}{R^2}}$$

The value of G can be found by rather careful experiments. It turns out to be $6{\cdot}7 \times 10^{-11}$ in M.K.S. units. The small size does not surprise us because it really measures the mutual force of attraction between two 1 kg masses placed 1 m apart, and we never observe ordinary-sized objects rushing together because of gravitational attraction!

2.25 Newton's successful guess

All inspired hunches have to be tested against observation and some years before Newton's work the mathematician Kepler had arrived at a connection between the time of orbit T of a planet round the sun and the radius R of its path. He did this by trial and error and painstaking calculations until he found an answer which fitted the data for all planets then known. The connection was that the ratio $\dfrac{T^2}{R^3}$ had the same value for each planet. Why was this ratio always the same? Kepler did not know the reason, but a simple use of Newton's formula explains the puzzle easily.

The centripetal force necessary for the planet's motion is supplied by the sun's gravitational pull – Fig. 2.37.

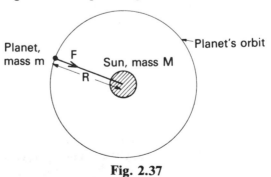

Fig. 2.37

$$\frac{mv^2}{R} = G\,\frac{Mm}{R^2}$$

which gives the orbital speed $v \quad = \sqrt{\dfrac{GM}{R}}$

The time for one orbit, $\qquad T = \dfrac{2\pi R}{v}$

$$= 2\pi R\,\sqrt{\frac{R}{GM}}$$

Squaring this gives: $\qquad T^2 = \dfrac{4\pi^2 R^3}{GM}$

or $\qquad \dfrac{T^2}{R^3} = \dfrac{4\pi^2}{GM}$

This will have the same value for all the planets and similar relations hold for all the satellites now in orbit round the earth, and for all the moons of Jupiter, and for any set of orbiting bodies having a common centre of attraction. The result is a good verification of Newton's law of gravitation.

3

Energy for Everybody

3.1 An ordinary word with a special meaning

'Eat Boffo Wheaty Chunks for breakfast and feel the energy difference!'

'If you haven't the energy take Vitaline tablets and go places.'

'Get your dog bouncing with Ener-Go, the energy packed dog food.'

We are bombarded from all quarters by jingles for this or that cereal or energy tablet and oddly enough they mostly use the word 'energy' almost exactly in its scientific meaning. The inference from the above mythical set of advertisements is that Boffo, Vitaline and Ener-go will enable the recipient to *do* something. They are supposed to give the ability to do some strenuous work and this is nearly the form of words physicists use for energy in the technical sense.

Here is a stone on the floor. Describe it to me. Well, it is about the size of half a brick, rough, muddy-tan in colour, quite hard to the touch, nearly five-sided apart from a missing corner, dry, probably porous, etc., etc. You could easily give a very detailed description of the stone. Now here is the same stone but on a shelf 2 metres higher than the floor. Describe it now. Still the same set of phrases apply to the raised stone, but although it appears to be just as it was there is a difference.

The stone on the shelf has more **energy** than the same stone on the floor because on the shelf it can do a job for us which on the floor it could not do. We could tie a string to it – Fig. 3.1 – pass the string over a pulley and drive a dynamo which in turn will light a lamp as the stone

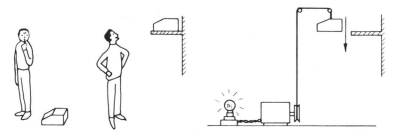

Fig. 3.1

falls. The stone has the ability to light the lamp only when the stone is raised off the floor. The stone has some **energy** 2 m up off the floor.

But, I hope you say, if we dig a hole two metres deep under the stone it will possess this energy again without having to be lifted! True, but we are discussing energy *differences* here. The stone at the bottom of the hole has less energy than the stone on the floor, which in turn has less energy than the stone on the shelf. On this reasoning the stone would have zero energy (of this kind) at the centre of the earth.

3.2 Potential energy

Why does the stone possess more energy higher up? Because of its position. Well, yes, but it falls down from the raised position because of the gravitational pull of the earth. This kind of 'position energy' is called **gravitational potential energy**, though often we leave out the first adjective and say merely 'potential energy'. Zero potential energy can be adopted for any convenient position and we will concern ourselves only with differences of potential energy from one position to another.

A reservoir of water is a good example of a stored quantity of potential energy. The larger the reservoir the more energy is stored and the potential energy is literally on tap at any time. Gravitational potential energy is the only kind of energy which can easily be stored in large quantities at present.

3.3 Moving objects – kinetic energy

Is there any other way in which the stone could set the dynamo spinning and light the lamp? If we sent it skidding along the table attached to a long string, it could do it when the string became taut – Fig. 3.2. The *moving* stone possesses the ability to light the bulb, the stationary stone does not, so we say the moving stone has some energy. The reason for its having energy is that it is moving and the name for 'movement energy' is **kinetic energy**.

Fig. 3.2

This name is applied to moving objects which spin as well as those which move bodily. Old-fashioned traction engines or steam road rollers used to have a large flywheel which would be set spinning rapidly when the machine had no great load to pull and which stored the energy as kinetic energy. The flywheel would help to pull the load when it became necessary. Modern friction-operated model cars are a small-scale version of the same thing – Fig. 3.3.

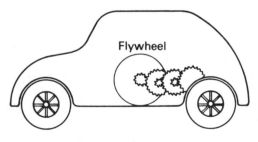

Flywheel

Fig. 3.3

3.4 Other forms of energy

Thinking of the stone again, it would not rise up to the shelf of its own accord nor go skidding along the table without being pushed. In other words it must get its energy from an outside agency. The kinetic or potential energy gained by the stone came from elsewhere.

Elastic or strain energy

If a spring were stretched and its end tied to the stone it could move the stone on being allowed to relax again. The stretched spring has energy to give to the stone – **elastic** or **strain** energy. Two forms of strain are shown in Fig. 3.4, though many clocks and simple toys make use of this form of energy.

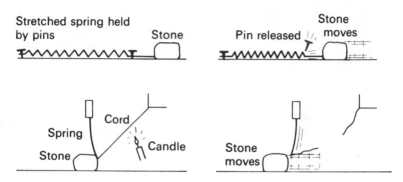

Fig. 3.4

Heat energy

If we had a small steam engine we could easily hoist the stone up the 2 metres to the shelf or, for that matter, drive the dynamo and lamp directly – Fig. 3.5. The hot steam does the pushing of the pistons in the engine and this is where the energy comes from – the steam possesses **heat** or **thermal** energy and also, because it is compressed steam, some strain energy.

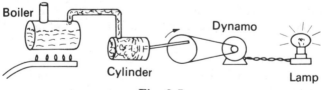

Fig. 3.5

The common uses of this kind of energy are met every day. Heat energy drives turbines to generate electricity and cook our food, keeps us warm and makes plants grow, is used in making metal joints and sterilises instruments in hospitals.

Chemical energy

The discussion of energy so far has merely used different adjectives to describe its various manifestations, we have not yet approached the ultimate source of energy. In the steam engine above, for example, the water is heated by gas, say, and the cold unburnt gas must have some energy in it which is released when it burns. **Chemical** energy is the term used for this and is the kind of energy possessed by all fuels and explosives – Fig. 3.6. More importantly it is also the form possessed by our foods which is, of course, where we get *our* energy from, energy to lift the stone up to the shelf or to maintain our body temperature at 36·9°C. Food is our body's fuel.

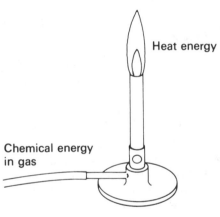

Heat energy

Chemical energy
in gas

Fig. 3.6

The most familiar example of a device using chemical energy is the internal combustion engine, that essential part of our civilisation.

Electrical energy

If you put certain chemicals in a box together you can build a cell which can light a lamp connected directly to it. The energy is fed from the chemicals to the lamp through conducting wires in the form of **electrical** energy – Fig. 3.7.

The fuel of the power station heats water to raise steam to drive turbines to turn dynamos to generate electrical energy. The simple bicycle dynamo does it from the kinetic energy of the bicycle. Some

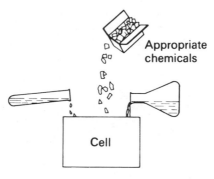

Fig. 3.7

cells can be 'refilled' with energy when their original supply runs out.

The role of electrical energy in our modern industrial society hardly needs any examples, the reader can easily supply his own.

Light energy
We are familiar nowadays with light-operated devices. Those doors which open miraculously just as you approach them, the ordinary photographic light-meter, solar cells carried by spacecraft, and on a different level the radar speed trap.

In the exposure meter used by photographers – Fig. 3.8 – the light falling on a sensitive part causes a needle to move in just the same way as electrical energy does. We say the energy this time is **light** energy.

Plants can use the light energy from the sun directly in the process called photosynthesis to make starch. Light energy can also affect the chemistry of a photographic emulsion or cause changes in the retinal cells of the eye to give us the sensation of sight.

Sound energy
In the days of supersonic airliners with sonic booms becoming increasingly common, no one will need reminding that **sound** energy is a real thing. Windows have been broken by sonic bangs and this is clearly an event requiring some energy – Fig. 3.9.

It is possible now to buy a lamp fitted with a box of electronics which switches the lamp on when someone claps their hands near by. Model trains can be bought which do the same thing. More commonly though sound energy appears at the end of the energy line rather than

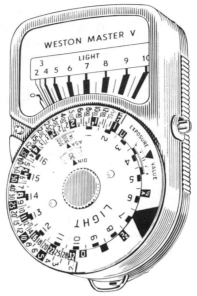

|Sangamo Weston Ltd.

Fig. 3.8 Exposure meter

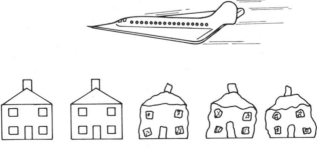

Fig. 3.9

as an initiating cause of other changes. No-one has seriously suggested a method of centrally heating a house utilising the noise made by its inhabitants, but the phrase 'hot air' has more than a grain of literal truth!

And the rest

The list is almost endless. If a new energy-possessing mechanism arises we coin another adjective. **Atomic** energy and **nuclear** energy are two twentieth-century names given to the capacity for energy release which the tiny atomic nucleus possesses. **Magnetic** energy and **radiation** energy are other forms which spring to mind.

The earth's supplier of energy

If we trace back all the sources of energy which we use in food, in the home, in industry or anywhere on earth, we ultimately find ourselves dealing with vegetation, dead or alive, and these of course derive their energy from the sun. We regard the sun as our prime energy supplier and, when you remember how far we are away from the sun and how little of the sun's rays we intercept, you realise there must be vast quantities of energy missing the earth and going off into space, apparently wasted at least to us.

Perhaps we are stretching it to say *all* our energy comes from the sun; atomic and nuclear energy are locked-up in the nuclei of atoms and molecules and do not need the sun's radiation to give them energy. Maybe though the sun and its planets had a common origin, so we might think that all the earth derives from the same source as the sun, whatever that was.

3.5 How do we measure energy?

We have talked about most of the ways energy can arise on earth, but we want to be able to measure it. What units shall we use? If we can find a suitable unit for one form of energy it will do for all the others as well. The easiest one to tackle is (gravitational) potential energy.

Suppose the stone we were thinking about earlier has a mass of 3 kg. On the earth, in Britain, it will experience a pull of gravity of 9·8 N kg^{-1} or 29·4 newtons in all. If you are to lift it up to the shelf steadily you will yourself have to apply a force of 29·4 newtons upwards. And you will have to exert this force over a vertical distance of 2 metres – Fig. 3.10.

We say you transfer 29·4 × 2 or 58·8 newton-metres of stored chemical energy in your body to the stone. The stone gains 58·8 newton-metres of (gravitational) potential energy. Another name for a newton-metre is a **joule** and this is what we use to measure energy. In

Fig. 3.10

this case 58·8 joules are involved, lost by the body and gained by the stone.

Generally we can write it down like this:

> **Energy transfer (joules) =**
> **Force (newtons) × Distance (metres)**

but we must be careful to measure the distance in the direction in which the force is acting. After all the stone cannot gain more potential energy by being lifted in a zig-zag way; if it is 2 m above the floor and weighs 29·4 N it has 58·8 J more energy on the shelf than on the floor, no matter how it got there.

One joule must be the energy transferred when a lump on which the earth exerts a force of 1 newton (i.e. a mass of $\frac{1}{9\cdot8}$ of a kilogram) is raised through 1 metre.

3.6 Energy and Work

In this example about the man and the stone, we see that the man's energy store goes down by 58·8 joules and the stone's goes up by 58·8 joules. All that has happened is that 58·8 joules have 'changed hands' between the man and the stone. Sometimes we talk about the man 'doing 58·8 joules of work' to lift the stone and this says the same thing again.

If you owe me £5, your store of money will go down, and mine up, by £5 when we settle the debt, just like the energy stores of the man and the stone. Most likely you will be very aware of the £5 actually leaving your hand as cash when you give it to me, and this corresponds to the 'work done' by the man.

Quite often the terms 'work done' and 'energy transfer' are used to describe changes, and either point of view can be taken. The simplest way to remember it, perhaps, is that work is done when energy is transferred from one thing to another and the work done equals the amount of energy transferred. Both would be measured in joules.

3.7 Potential and kinetic energy formulas

We have decided to use potential energy to make our unit of energy – the joule. Let us put it in the form of a formula now. We have said:

$$\text{Energy transfer} = \text{Force} \times \text{Distance}.$$

If a mass m kg is placed in a gravitational field of strength g newtons per kilogram the force on it is mg newtons. Let this be lifted through h metres.

Energy transfer = Work done = mgh (joules)

This gives the answer 58·8 again if we try the numbers from Section 3.5.

When we deal with kinetic energy we use the simple case of the stone falling freely for the 2 metres through which it was previously raised. Let us work this through in numbers and symbols.

The stone falls downwards under the pull of gravity for 2 metres. Just before it hits the floor it is moving fast and the potential energy it had on the shelf is now its kinetic energy – Fig. 3.11. We want to find the connection between the mgh of potential energy and its velocity just as it reaches the floor.

The earth's field strength is 9·8 N kg^{-1} which means that the stone will accelerate as it falls at 9·8 m s^{-2} (Section 2.9). How long will it take to fall the 2 metres to the floor? Using the formula $s = \frac{1}{2}at^2$ from Section 2.3 we have:

$$2 = \frac{1}{2} \times 9 \cdot 8 \times t^2$$

or

$$t = \sqrt{\frac{4}{9 \cdot 8}} \text{ seconds.}$$

$$t = \sqrt{\frac{2h}{g}}$$

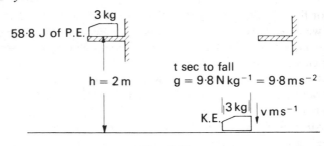

Fig. 3.11

The velocity v with which it hits the ground comes from $v = at$ (Section 2.3).

$$v = 9{\cdot}8 \times t$$

$$= 9{\cdot}8 \sqrt{\frac{4}{9{\cdot}8}} \text{ metres per second} \qquad v = g\sqrt{\frac{2h}{g}}$$

$$= \sqrt{4 \times 9{\cdot}8} \text{ metres per second} \qquad = \sqrt{2gh}$$

or $\qquad v^2 = 4 \times 9{\cdot}8\ (\mathrm{m\,s^{-1}})^2 \qquad\qquad\quad v^2 = 2gh$

We know that the kinetic energy at the bottom must equal the potential energy at the top, and that we know was 58·8 J or *mgh*.

So, kinetic energy $= mgh$

$$= mg\,\frac{v^2}{2g}\ \text{from the last line of algebra above.}$$

or $\qquad \underline{\text{KE} = \tfrac{1}{2}mv^2}$

This result can now be used for any moving object to work out its kinetic energy

$$\boxed{\begin{array}{c}\textbf{Kinetic energy} \\ \textbf{(joules)}\end{array} = \tfrac{1}{2} \times \begin{array}{c}\textbf{Mass} \\ \textbf{(kg)}\end{array} \times \left(\begin{array}{c}\textbf{Velocity} \\ \textbf{(ms}^{-1}\textbf{)}\end{array}\right)^2}$$

3.8 Conservation of energy

You will have noticed in the last section that we decided to say the KE at the bottom of the 2 metre fall *equals* the PE at the top, giving us

$\frac{1}{2}mv^2$ for the KE. Provided we use $\frac{1}{2}mv^2$ for the Kinetic Energy of an object we make sure that the total of (PE + KE) is always the same. There is no point in trying to test this by experiment – we measure kinetic energy in such a way that it must be so.

Later on we need to measure other forms of energy and these also are worked out in such a way that the total amount of energy in a system is always constant unless an external supply is available. This is often called the Principle of the Conservation of Energy and it is an extremely useful idea in physics, but it is not a startling result – we make it true by measuring energy as we do.

On the other hand, the Conservation of Momentum mentioned in Chapter 2 *is* a startling result since it tells us something very important about the way matter behaves when collisions occur – namely that each object in a collision exerts the same force on the other.

3.9 Power – the Watt

To someone concerned with traffic flow it is not much help to know that 29 006 325 vehicles have passed a certain point since counting began some years ago. What the traffic engineer usually wants to know is how many vehicles per hour – the *rate* at which they pass each road junction. In the same way the important thing about a large river is not how much water it contains, but at what *rate* the water is flowing.

When an energy transfer is occurring, say, in an electric kettle element, the quantity most useful to know is at what *rate* the electrical energy is being transferred into heat energy, in other words how many joules per second are involved. As a convenient shorthand we call a joule per second a watt.

1 watt = 1 joule per second.

This means that a 1500-watt kettle converts energy at a rate of 1500 joules every second. The output of power stations is often quoted in megawatts – millions of joules per second. 40 watts of sound energy will fill a fair-sized hall, but you would need 2000 watts of heat energy to heat an ordinary room.

We use the watt as a unit of measurement very freely, more so probably than the joule, and this often leads to the cart being put before the horse –

$$1 \text{ watt-second} = 1 \text{ joule}$$

(a watt-second, not a watt per second).

When you pay the electricity company you are paying for the total amount of electrical energy you have converted into other forms every three months – i.e. you pay for *joules* – and the electricity meter – Fig. 3.12 – is really a joule-meter. However, it comes on your bill as so many 'kilowatt-hours' and not so many joules.

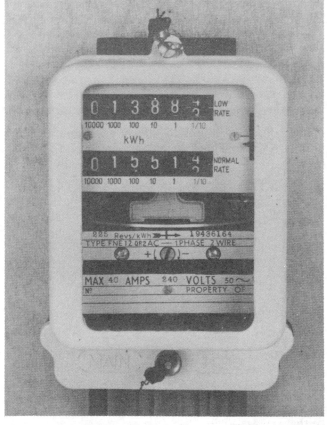

The Electricity Council

Fig. 3.12   A kWh meter

$$1 \text{ watt-second} = 1 \text{ joule}$$
$$1 \text{ watt-minute} = 60 \text{ joules}$$
$$1 \text{ watt-hour} = 3600 \text{ joules}$$
$$1 \text{ kilowatt-hour} = 3\,600\,000 \text{ joules}$$

At the rate of $3\,600\,000$ joules to the electricity board's 'unit' you get through quite a few joules in a quarter – and that has not included the gas bill!

If you are thinking of economising on electricity look for the wattage each appliance is rated at, usually stamped somewhere on the back or underneath. The high power ratings are the money-eaters: immersion heaters, convector or radiant fires, etc. Ordinary electric lamps, filament or fluorescent, and electronic gadgets like tape-recorders or televisions have a very low power rating and are not worth bothering about if you compare their running cost with other large power items.

3.10 Measuring electrical energy

Electrical appliances are generally rated in watts as we have just discussed, yet the watt is basically a mechanical unit – it comes from the joule which in turn derives from the newton, both mechanical measures. How then does the watt turn up in electricity?

The answer lies in the volt and the amp which will be dealt with more fully in Chapter 7. Because of the way in which these electrical units are defined, the connection between them and the watt is very simple.

$$\frac{\text{Power}}{\text{(watts)}} = \frac{\text{Current}}{\text{(amps)}} \times \frac{\text{Potential difference}}{\text{(volts)}}$$

or in the symbols commonly used for these quantities:

$$\boxed{W = IV}$$

This is a matter of definition, not a thing to be tested experimentally. Why are there 60 seconds in a minute? Because that is what a minute is, we have decided it. Why do watts equal amps multiplied by volts? Because that is what a volt is, we have decided it: a volt *is* a watt per ampere (see also Section 7.3).

It is an easy matter now to measure electrical energy even if there is no joulemeter available, all we need is an ammeter, a voltmeter and a clock. If a soldering iron takes a $\frac{1}{10}$ amp current from the 240 volt mains we know it uses energy at a rate of $\frac{1}{10} \times 240$ watts, or 24 watts. How many joules in 5 minutes? $24 \times 5 \times 60$ joules will be the number, that is 7200 joules.

The starter motor on a car may well take a current of 200 amps from the 12 volt battery on a cold morning. That means 2400 watts. If it takes 5 seconds to start the car the energy needed will be 12 000 joules! This is clearly a very rapid way to exhaust the energy stored in the battery.

3.11 Measuring heat energy – specific heat capacity

Let us take a small immersion heater, Fig. 3.13, and put it in a bowl containing 1 kg of water. We connect it through a joulemeter, Fig. 3.14, or an ammeter and voltmeter, so that we can measure how much electrical energy is supplied to the water. The thermometer will give us the rise in temperature. Here are some typical figures.

Using a heater for 5 minutes in 1 kg of water the rise in temperature is 2·4 degrees C, and 12 000 joules are clocked up on the meter. To give a rise of only 1 degree C would need $\dfrac{12\,000}{2\cdot4}$ joules or 5000 joules.

Similar experiments using more or less water show that this figure per kilogram per degree rise is always the same for water, at least within the accuracy of the measurements.

Suppose we heat a kilogram of something else, say aluminium or cooking oil, with the same heater for the same time under the same conditions, will there be the same temperature rise? Experiments say no. With aluminium the rise is about 12 degrees, five times as much as water. Working it out as we did for water, aluminium requires only 1000 joules for every kilogram and every degree rise. It is five times as 'heat-up-able' as water, or it needs only $\frac{1}{5}$ as much heat energy mass for mass to get it equally hot.

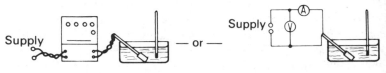

Fig. 3.13

Philip Harris Ltd

Fig. 3.14

The numbers 5000 and 1000 are rough values for what we call the **specific heat capacity** of water and aluminium respectively, measured in joules per kilogram per degree C. Water has an unusually large specific heat capacity. It is more difficult to heat up than most other substances, a property which is explained by the complex molecular arrangement of water in the liquid state and which is accompanied by abnormally high density and boiling point compared with other common liquids.

3.12 Energy gets 'lost'

5000 and 1000 J kg^{-1} $C^{\circ-1}$ are only rough values for the specific heat capacities of water and aluminium. In the list of accepted measurements you will find the numbers 4180 and 870. Why does the simple measurement described in the last section go wrong? The error is always the same way, it seems to require *more* joules to heat up *our* water than that used by the 'proper' scientists!

The reason is that we have not allowed for any of the supplied energy going elsewhere than into the water. The container will take up some joules of energy and, as soon as the water is hotter than the air around it, heat will be given to the atmosphere. Nor might we have stirred the water thoroughly nor recorded the highest temperature it reached. All these reasons will make our result too high, and we say loosely that energy was 'lost'. It is lost, of course, only as far as getting to the thermometer was concerned, but taking the universal view those joules will be somewhere making the earth or its atmosphere slightly warmer.

Many different kinds of experiments on energy transfer have been tried and always the experimenter has to take elaborate care to allow for or avoid any energy escaping from the system he is using. The most usual way this happens is as heat energy due to friction where two materials rub on each other, and this energy is virtually impossible to measure. All hot objects transfer heat to cooler ones and this is another source of 'lost' energy, but one which is much easier to make allowance for.

3.13 Energy transformations

The story of physics and of a large part of engineering is contained in the desire to change the energy from a given source from one form into a more convenient form. This is what machines are designed to do and so is the human body.

Somehow we have to maintain our body temperature about 22 degrees C above our surroundings and we must be able to regulate the rate at which we lose energy (by shivering or sweating, for example) so that we stay at an even temperature under all conditions. We need to have a furnace inside to supply the energy and this requires fuel to keep it going. We can transform the chemical energy in foods or body fat into heat energy to keep us warm, or into sound energy to enable us to talk, or to kinetic and potential energies when we move or climb stairs.

The internal combustion engine is a familiar device for transforming the chemical energy of a fuel into kinetic energy and heat energy.

A power station, Fig. 3.15, burns a fuel to heat water to raise steam to turn turbines to drive alternators to generate electrical energy, a long and complicated chain of energy transformations. The length of the chain is the main reason why power stations can achieve overall efficiencies of around a mere 25 % for at every transformation along the way some energy will be 'lost' or become unusable. See also Section 3.15.

Central Electricity Generating Board

Fig. 3.15 Bradwell nuclear power station

A microphone transforms sound energy into electrical energy and a loud-speaker makes the reverse transformation. A thermopile might be able to detect the heat from a candle at 2 km distance by transforming its radiation energy directly into electrical energy. During the electrolysis process widely used in chemical engineering electrical energy is changed into chemical energy. Fuel cells, as used in space capsules, perform the opposite of this and produce electrical energy directly from the interaction of chemicals.

3.14 Machines – what they can and cannot do

My car weighs about 1000 kg and with the aid of a screw jack I can hoist one corner off the ground without a great deal of effort, something I certainly could not do with my bare hands. The screw jack seems to give me the ability to do a job I could not do otherwise. According to the description of energy in Section 3.1, it appears that I have gained some by using the jack.

Unfortunately this is not so! Energy does not come free like this. In fact in terms of joules I use more lifting with the jack than with my bare hands. The reason for this is that some energy is needed to operate the machine. All passive arrangements of gears, wheels, pulleys or levers are similar in this respect – they will magnify a force easily but not energy.

The simplest kind of machine is a lever, a can-opener is a good example – Fig. 3.16. The pivot is much nearer the pointed end than the handle and if the distances are 12 mm and 84 mm any force applied upwards at the handle end should appear magnified seven times at the pointed end. Of course, the handle end has to move seven times as far as the pointed end and, since the energy transfer is measured by the product of force multiplied by distance moved (Section 3.5), we get out only as much energy as we put in. In fact some energy is probably taken at the pivot, so we will not even get as much out as we put in.

Machines in which there are moving parts, like gears or pulleys, suffer badly in this respect because the parts in contact have frictional forces acting when they move and these account for a good proportion of the energy input since work has to be done to overcome these forces. Perhaps only 40, 50 or 60% of the input energy will be available as useful energy for output.

Consider the pulley block system of Fig. 3.17. If the load is raised

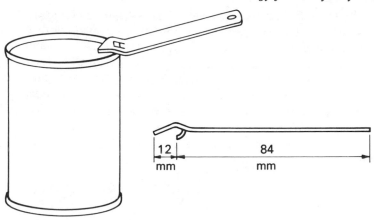

Fig. 3.16

1 metre, there will be six metres of rope to haul in since the load is supported by six lengths of rope. Ideally, then, if the load is 120 newtons, there will have to be 120 joules of energy transfer to raise it 1 metre. Having 6 metres to pull in at the other end would suggest a force of only 20 newtons required. Actually it is likely to be 30 or 40 newtons because the lower pulley block must be lifted with the load and each pulley will need some turning even if well-oiled. The efficiency of the machine will be well below 100% and machines designed to cope with heavy loads (such as the block and tackle used to lift the engine out of a car) are deliberately designed to be less than 50% efficient.

Suppose the load and lower block of Fig. 3.17 together weigh 150 N (i.e. their combined mass is such that the earth's gravitational pull on them is 150 N), and the force needed to lift them is 60 N. When the load rises 1 m the energy transferred to it is 150 J. The energy transferred into the machine by whoever pulls is 60 × 6 or 360 J, so that 210 J are needed to overcome the frictional forces and get the pulleys moving for each metre rise of the load. If the 60 N pull is now removed, what happens? The load and block could only transfer 150 J of energy if they fell 1 m and that is not sufficient to move the pulleys, so the machine does not run backwards if the person pulling suddenly lets go. For a heavy load this is an important safety feature. The reader might like to check that a machine like this will not run backwards if its efficiency (the ratio of useful energy output to energy input) is 50 % or less. In the example here the efficiency is 41·7%.

Devices which transform energy from one kind into another always involve some energy loss. If an electric current flows through a wire heat is developed which cannot·be usefully exchanged for other kinds of energy. Noisy machines involve sound energy which is virtually irretrievable. Where materials twist or bend energy is involved which usually ends up as heat. The tyres of a motor car are a good example of this.

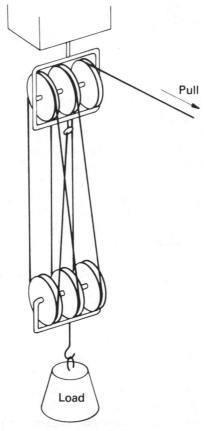

Pull

Load

Fig. 3.17

No passive machine ever invented or likely to be invented can turn out more energy than it was originally given. Devices like power-assisted brakes do not do so, of course, because they draw energy from a battery or dynamo or other source of energy, as do electronic amplifiers.

3.15 Heat engines – things are worse than we thought

From what was said in the last section there does seem a reasonable future for machines in the energy field. We lose some energy, yes, but the gain in convenience more than outweighs that disadvantage and we are prepared to pay for our pleasures. We do some careful designing, use the oil can and reduce our losses to the minimum. The transformer (Section 7.49) is probably the commonest machine coming nearest to perfection and that could come nearer if we were prepared to pay for it. Typically about 93 % of the energy fed in is available at the output side as useful energy – a magnificent figure.

Yet there is one kind of energy conversion which offers no hope of high efficiency and perversely we use it perhaps more than any other. Our heritage of fuels in the earth, wood, peat, coal, oil, gas, has always put a premium on methods of converting heat energy into other forms. Unfortunately heat energy is the only kind which cannot be so converted without losing a high proportion of it.

An analogy may help to make the problem clearer. (This analogy is only *like* the heat case, not the same as it in *all* respects.) Consider a reservoir of water at the top of a hill linked to a turbogenerator at the bottom – Fig. 3.18. Water flows through a wide pipe down the hill and turns the turbine blades to generate electricity. Think of the energy changes. Potential energy of the water in the reservoir becomes kinetic energy in the moving water, then kinetic energy in the turbine blades and eventually electrical energy from the generator. This is an easy enough energy chain to follow, but is not the end of the story because the water still has to get away from the power station and it takes with it some of the kinetic energy it arrived with.

We are doomed always to 'throw away' this energy, because if we try to use it by putting another turbine downstream this will slow the water down and react back on the original one. Any attempt to make use of the kinetic energy of the escaping water causes a slowing down of the turbine and it turns out that we lose more than we gain. There is

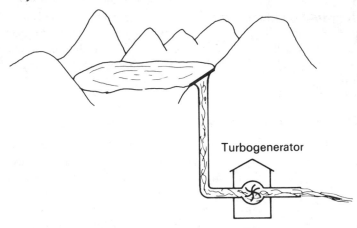

Fig. 3.18

no escape from this – the water must be moved away or the whole thing stops. The best we can do is to build the reservoir as high as possible.

The main sequence here can be represented by the diagram of Fig. 3.19. Clearly, in the absence of energy losses, W, the useful energy output is:

$$W = Q_1 - Q_2$$

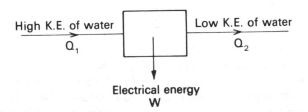

Fig. 3.19

and the overall efficiency is:

$$\frac{\text{Useful energy output}}{\text{Total energy input}} = \frac{W}{Q_1}$$

$$= \frac{Q_1 - Q_2}{Q_1}$$

$$= 1 - \frac{Q_2}{Q_1}$$

This expression is always less than 1 as long as Q_2 is there and, as we have seen, there is no hope of avoiding Q_2.

Heat conversions are like this. Heat energy is taken, for example, from the hot gases inside the cylinder of a petrol engine and is used to give kinetic energy to the piston and the rest of the car. The waste gases then have to be removed and they take with them a fair proportion of the original energy of the petrol. Any attempt to put a device in the exhaust pipe of a car will have predictable results!

Power stations carry the same burden. What a waste to see all that energy escaping from cooling towers. Yet any attempt to use that energy would react back on the power station and make it less efficient still. We are forced to throw away large quantities of unavailable energy. One possible end of the universe may arise when all the heat energy in it is in this unavailable form!

3.16 Atomic energy

This chapter on energy must contain a reference to what may become a common source of energy in the future. Power stations in various parts of the country already use uranium as a fuel instead of coal. It does not burn like coal or oil, but it certainly gets hot and the sequence of energy changes, fuel – heat – steam – turbine – generator is still the one used.

Where does uranium get its fuel capacity from? The answer is that the nucleus of the uranium atom can be made to split up into smaller pieces and release energy in the process (see Chapter 9). Because quite ordinary sized pieces of uranium contain millions of atoms, large amounts of energy are available from comparatively small quantities of fuel. The energy is released as heat which is used to raise steam, etc. The process is controlled by slowing down or speeding up the rate

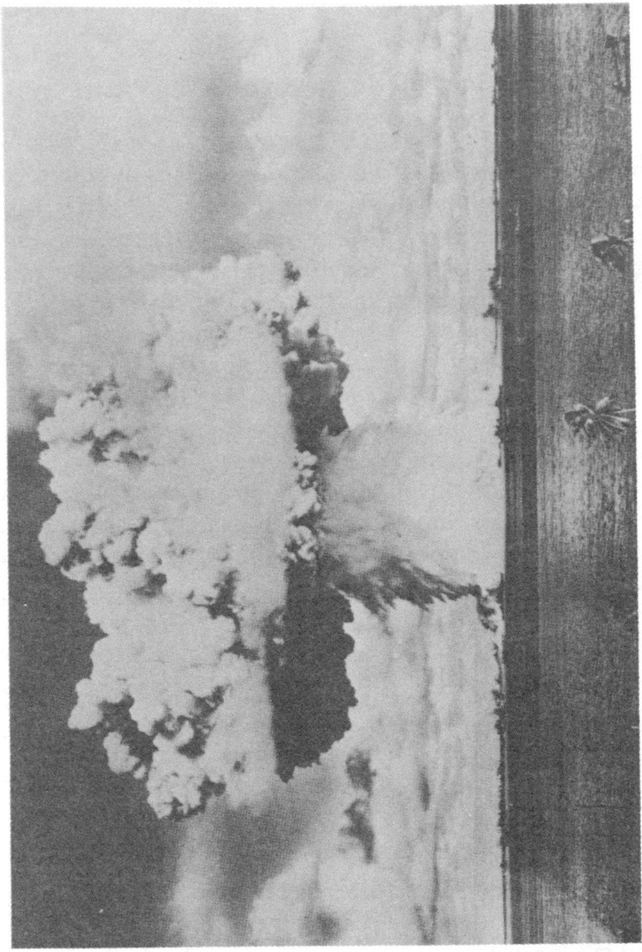

Fig. 3.20 Bikini Atoll atomic bomb explosion, 1946

at which uranium atoms split. The atomic bombs which ended the Second World War in Japan are warnings of what might happen when the splitting process (called *fission*) is allowed to go unchecked – Fig. 3.20. The subtle point is that each fission can trigger off others and we soon have an uncontrollable chain reaction unless elaborate steps are taken to avoid things getting out of hand. The reactor in a nuclear power station is really an atomic bomb in which fission occurs at a regular and controlled rate.

3.17 The sun's energy

A star like the sun which can keep shining for many thousands of millions of years must have great energy resources somewhere. It is, in fact, a gigantic furnace where instead of atomic nuclei being broken down by the fission process small nuclei are being built up, or fused together, to make larger ones. The energy release here is much greater than the fission process involves. Man's only successful attempt so far to copy the idea is the hydrogen bomb.

The trouble with the *fusion* process is that it needs a temperature of millions of degrees before it will go, and what earthly way is there of containing any material as hot as that? So far the 'bottle' which will hold a fusion reaction has not been invented, but there is hope that it will be. The raw material for fusion is hydrogen and we have plenty of that in water in the oceans of the world. It is to be hoped that some way *is* found to make a controlled sun; the energy supply problems of the world would then be solved for good!

4

Gas Properties – How Cold Can You Get?

4.1 What we are looking for

When we were discussing the molecular model of materials in Chapter 1 we described a gas as being the most disorganised state – molecules far from each other travelling at high speed and bouncing into each other and the sides of the vessel containing them. Armed with a little mathematics, in Chapter 2 we were able to work out roughly how fast the molecules must be moving and obtained an answer of realistic size.

This chapter on gases is concerned with their bulk properties. Here the effects of individual molecules are blurred over and only their average behaviour is important. When dealing with comparisons between countries their relative economic well-being is gauged from the overall performances in exports and industrial production, and the effect of a single worker or group of workers is much too small to be seen. In the same way here we want to know how large numbers of molecules behave.

The bulk properties of a gas are its pressure (force per unit area), its volume and its temperature, which we will symbolise as p, V and T respectively. A common experience when pumping up a bicycle tyre is that when air is squashed it gets hot, i.e. T goes up when p goes up and V goes down. Exactly how are these three quantities linked together? We shall deal only with a fixed mass of gas and consider how *two* of the quantities are related when the third one is not allowed to alter.

4.2 Pressure and volume – keeping temperature fixed

To find this connection we need to have a trapped quantity of a gas, air will do, and alter the pressure on it using a pump. The corresponding values of volume are easily measured if the air is trapped in a graduated tube. Fig. 4.1 shows a simple way of arranging things to find out about pressure and volume.

The pressure is measured directly on the gauge in newtons per square metre, and the volume read off the scale behind the tube. A pump can be used to compress the air initially and subsequent readings taken by releasing the pressure bit by bit through the tap. The only serious precaution necessary is patience since we require the temperature to remain the same through all the readings. Time must be allowed between each adjustment to allow the air in the tube to

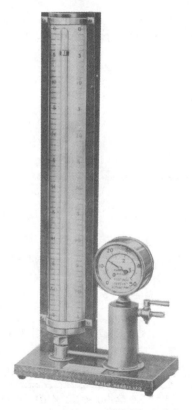

Philip Harris Ltd

Fig. 4.1

regain the surrounding temperature. Another reason for waiting a little is to allow the oil in the tube to run down the sides before the volume figure is recorded.

The results of such an experiment give a graph like Fig. 4.2. Another graph can be drawn like Fig. 4.3 using $\dfrac{1}{\text{pressure}}$ instead of pressure, or again the product of pressure × volume can be calculated for each case as shown in Fig. 4.4. There is no doubt that these figures follow a pattern giving the same value of $p \times V$ each time. Tests with other gases show a similar behaviour and the connection between pressure and volume can be stated like this:

For a fixed mass of gas the pressure and volume are inversely proportional to each other provided the temperature remains steady.

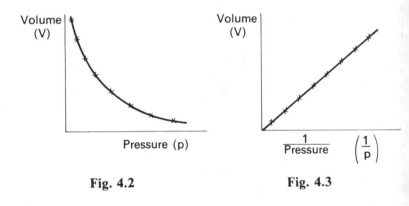

Fig. 4.2   Fig. 4.3

Pressure (p) N m^{-2} (×10^5)	2·62	2·27	1·93	1·71	1·52	1·35	1·17	1·04
Volume (V) m^3 (×10^{-6})	17·2	19·9	23·9	27·0	30·0	35·0	40·0	44·3
pV joules	4·50	4·50	4·62	4·61	4·56	4·72	4·68	4·60

Fig. 4.4

In symbols we can write:

$$pV = \text{a constant}$$

$$\text{or} \quad p \propto \frac{1}{V} \text{ and } V \propto \frac{1}{p}$$

All these three statements mean the same thing – if we halve the pressure the volume doubles; if we increase the pressure by a factor of 5 the volume becomes $\frac{1}{5}$ of its former value.

This is a celebrated result called **Boyle's Law**, after Robert Boyle who first investigated the problem systematically in 1662.

What does the law mean in terms of molecular behaviour? Pressure is caused by collisions between molecules and unit area of the containing vessel and depends on the velocity of impact, the mass of the molecules and the frequency of collisions (Section 2.18). If any of these factors is altered the pressure will alter and in this case we are not giving the molecules any extra speed nor changing their mass. We are giving them more room in which to move and so increasing the time between collisions with the walls, causing fewer collisions per second and a lower pressure. Squashing them into a smaller space conversely gives a higher pressure.

4.3 Pressure and temperature – keeping volume fixed

This is much the easiest relation to find experimentally. A flask, a pressure gauge, a water heating bath and a thermometer are required. (*NB* There is a snag here because we have not yet decided what temperature *is* apart from the notion of 'hotness', but for the time being a mercury-in-glass thermometer will be used and the whole problem discussed at length in Section 4.10.) The procedure is simple: Fig. 4.5 shows the layout, and readings of pressure and temperature are taken at intervals between the freezing and boiling points of water. A flask with quite large volume is needed in order to render unimportant the unheated space in the connecting tube and pressure gauge. Good stirring is also needed. Typical figures are shown in Fig. 4.6.

The results this time lie near to a straight line for pressure and temperature, as sketched in Fig. 4.7. The two quantities are *not* directly proportional to each other because the line does not go through the zero point of both scales, but if we draw the line backwards to where it meets the temperature axis we *could* get direct proportion by calling that point the zero of temperature – Fig. 4.8. It comes at about $-273\,°C$.

This is just a mathematical trick at the moment to get the quantities in direct proportion. If we were to try to cool the gas as far as this, it

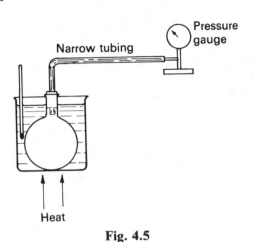

Fig. 4.5

Pressure (p) N m^{-2} ($\times 10^5$)	0·94	0·98	1·01	1·09	1·12	1·23	1·29
Temperature (t) °C	0	12	20	40	50	80	100

Fig. 4.6

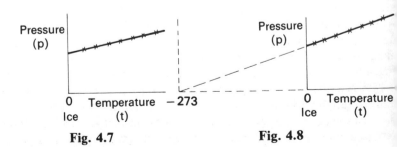

Fig. 4.7

Fig. 4.8

would first liquefy and then freeze solid and would not be a gas at all. However, $-273\,°C$ would be that temperature at which the pressure of the gas becomes zero, if the gas continued behaving as it does at ordinary temperatures.

The result can be put into words like this:

For a fixed mass of gas whose volume is kept steady, the pressure increases by equal amounts for every degree rise in temperature and at $-273\,°C$ its pressure would be zero.

In symbols: $p \propto (t + 273)$, if t is the centigrade temperature,

or $\dfrac{p}{(t + 273)} =$ a constant value.

What is happening to the molecules? Of the factors which decide pressure it is not a matter of more space or a different mass. We give them more *energy* as we heat them, the energy coming from whatever kind of heater is used to get the water bath hot. The energy will be kinetic energy of the molecules of the gas, so we are really altering the *speed* of the molecules and thus producing a larger impulse whenever they collide with the walls of the vessel. (*Note:* this may only be an *average* increase in energy among the molecules. In fact when dealing with bulk effects like pressure we have no means of knowing whether all or only some of the molecules are travelling more rapidly than when they were cool, but we certainly supply energy to them as a whole, so on the average they must speed up.)

4.4 Volume and temperature – keeping pressure fixed

This time we wish to allow the trapped air to expand if it must, but always in such a way that its pressure remains the same. A fairly simple way to do this is to use atmospheric pressure as the reference point and let the sample of air always come to equilibrium with this. A flask can be fitted with a capillary tube which contains a short thread of mercury sealing the air into the flask – Fig. 4.9. The mercury will be pushed up or down each time the pressure inside the flask alters, but it will always leave the pressure in the flask equal to the outside pressure.

The position of the mercury thread gives the volume of the gas if the flask's volume up to the bung is known and, again, a mercury-in-glass

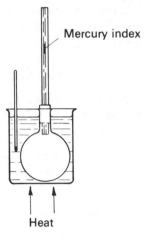

Mercury index

Heat

Fig. 4.9

thermometer gives the temperature. A set of results will produce a graph similar to the pressure graph of the last section and we can try the same trick with this. If we produce the line backwards and call the temperature it reaches zero, we can say that the volume of a gas and that sort of temperature vary directly with each other – Figs. 4.10 and 4.11.

The astonishing thing is that this graph again hits the temperature axis at $-273\,°C$, or very nearly. Even more surprising is that when the pressure experiment or the volume experiment is repeated with any of the common gases, oxygen, nitrogen, hydrogen, etc., they *all* give a line

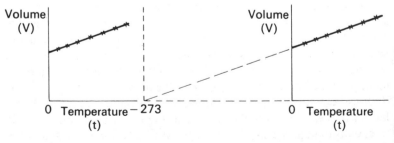

Fig. 4.10   **Fig. 4.11**

which leads backwards to the same point of $-273\,°C$. Clearly there is something rather special about that temperature. It gives us a new zero for measuring hotness which allows us to say that pressure and volume are each directly proportional to temperature starting at $-273\,°C$.

For a fixed mass of gas whose pressure remains steady the volume increases by equal amounts for every degree rise in temperature and the volume would become zero at $-273\,°C$.

In symbols: $\qquad V \propto (t + 273)$

$\qquad$ or $\qquad \dfrac{V}{(t + 273)} = $ a constant value.

On the molecular level things are quite interesting. Once again we are supplying energy which makes the molecules move about faster as the temperature rises, but to prevent this effect causing an increase in pressure the space occupied by the molecules goes up, reducing the frequency of collisions. The two effects work against each other to keep the pressure unchanged.

4.5 Absolute temperature – the Kelvin

We have seen that if we shift the place from which we measure temperature from the normal $0\,°C$ at the ice point to a place 273 Celsius degrees colder, we can get simple proportionality between it and the bulk properties of pressure and volume. In fact this is also true of the product (pressure × volume). For rather complicated other reasons we believe this new zero temperature to be as cold as it is possible to get. Obviously gases could not have negative volumes and pressures, but in terms of molecular motion it means the temperature at which the molecules are moving as slowly as possible. (Advanced physics shows that even at this absolute zero the molecules do have small amounts of vibrational energy.)

Measuring temperature from this zero gives us a new number 273 larger than before for every ordinary temperature. (There are still 100 degrees between the freezing and boiling points of water, so we still have a *centigrade* size of degree.) Thus ice melts at 273 K, body temperature is about 310 K and water boils at 373 K, the symbol K indicating this new temperature, called *Kelvin*. The scale which starts at zero K is known as the **absolute scale of temperature** and its zero is the coldest possible cold.

No method of cooling yet devised can get down to this absolute zero of temperature nor will it ever be possible. The reason is that all methods of cooling extract only a fraction of the heat energy in a material and this means we can never get it all out. Suppose a cooling process takes $\frac{1}{2}$ the heat energy in one operation. Repeating the process will remove $\frac{1}{2}$ of the remainder leaving $\frac{1}{4}$ behind. Next time $\frac{1}{8}$ is left, then $\frac{1}{16}$ and so on, but it never gets down to nought.

Low temperature physicists can reach within a small fraction of a degree of the absolute zero (which is actually at $-273 \cdot 15\,^\circ\text{C}$), but it gets harder and harder to go any lower and more and more expensive on equipment.

4.6 A combined gas law

Looking back at the results of the last three sections we can collect them together using the symbols as before:

For a fixed mass of gas
$$\begin{cases} pV = \text{a constant if } t \text{ is not changed,} \\[2mm] \dfrac{p}{(t+273)} = \text{a constant if } V \text{ is not changed,} \\[2mm] \dfrac{V}{(t+273)} = \text{a constant if } p \text{ is not changed.} \end{cases}$$

These three relations can be put together into one:

$$\frac{pV}{(t+273)} = \text{a constant (depending only on the mass of gas and what gas it is)}$$

or
$$\frac{pV}{T} = \text{a constant, } R$$

or finally
$$\boxed{pV = RT}$$

where T is the absolute temperature and R has a value determined by the quantity and nature of the gas.

4.7 How big is R?

All we need to find the size of R are values for each of p, V and T. A fair-sized room might be 3 m high with a floor measuring 4 m × 5 m, i.e.

of volume $60 \, m^3$. Normal atmospheric pressure is very near $100\,000 \, Nm^{-2}$ (Section 2.18) and room temperature can be taken as near enough 300 K. Putting in these values gives for the 75 kg of air in the room,

$$R = \frac{pV}{T}$$

$$= \frac{100\,000 \times 60}{300 \times 75} \, J \, kg^{-1} \, K^{-1}$$

$$= \underline{267 \, J \, kg^{-1} \, K^{-1}}$$

The value worked out like this will be correct only for air. What about other gases? They will have a different R, at least if we work it out per kilogram they will, but there is a very important property of gases well known to chemists which we can use to obtain a value of R which will be the same for all gases. Physical chemists often work with a special amount of gas which they call a 'mole' of gas. This amount contains a certain *number* of molecules, not a certain *mass*, and according to an important chemical principle equal volumes of gases at the same temperature and pressure contain equal numbers of molecules. Consequently 1 mole of any gas will have the same volume at a certain temperature and pressure. For instance, at 273 K and $100\,000 \, N \, m^{-2}$ that volume is $0.0224 \, m^3$. These numbers can give the special value for R which is the same for all gases:

$$R = \frac{100\,000 \times 0.0224}{273} \, J \, mol^{-1} \, K^{-1}$$

$$= \underline{8.22 \, J \, mol^{-1} \, K^{-1}}$$

4.8 Compare $pV = RT$ with $pV = \frac{1}{3} nmv^2$

We are now in a position to look at the result of the last section (which comes from the observed bulk behaviour of gases) alongside the similar result of Section 2.17 (dealing with molecular ideas). The two equations are:

$$pV = RT \text{ and } pV = \tfrac{1}{3} nmv^2$$

The second equation came from purely theoretical work and if the theory it was based on has any validity it ought to explain the experimental foundation to the first equation.

Such an explanation is possible by interpreting the idea of temperature in a physical way. If we take absolute temperature to be proportional to the average kinetic energy of the molecules, theory and experiment fit very well. From Section 3.7, kinetic energy is $\frac{1}{2}mv^2$, so if we write:

$$\frac{1}{2}mv^2 = kT$$

where k is a linking factor between temperature and energy, we go on to put:

$$pV = \frac{2}{3}nkT$$

This simple interpretation of temperature yields surprising dividends:

(*a*) If temperature T and mass (i.e. n) remain fixed, pV = a constant.

(*b*) If pressure p and mass remain fixed, $\dfrac{V}{T}$ = a constant.

(*c*) If volume V and mass remain fixed, $\dfrac{p}{T}$ = a constant.

(*d*) If volume, pressure and temperature are fixed, so is the number of molecules n.

The first three of these deductions are the results of Sections 4.2, 4.3 and 4.4, and the last one is the chemical principle referred to in the last section which is known as Avogadro's hypothesis.

4.9 Real gases are not as perfect as all that!

The agreement between the bulk behaviour of gases and the predictions of the molecular theory seems at first sight to be almost too good to be true; rarely can we tie up any branch of physics quite as neatly as that! The agreement is certainly close, but when accurate experiments are performed we see it is not as close as we would like it.

For instance, real gases do not give a perfect fit to Boyle's Law (Section 4.2). All gases show small deviations from the simple rule, some more severe than others. The ones which fit best are hydrogen and helium, the worst being those gases which can easily be liquefied or frozen like carbon dioxide or methane. Why do real gases not quite fit the 'perfect' equation $pV = RT$?

This is quite a complicated problem, but there are two things which clearly must make a difference. When the simple molecular theory of Chapter 2 was worked out, two assumptions were made to help the arithmetic along which really must be questioned now. The first was that the molecules had all the volume V of the rectangular box in which to move, no account being taken of the size of the molecules themselves. For a large box and few molecules this is perfectly acceptable, but when the pressure gets large so that molecules are closely packed this could become an important error. Our measured volume V will be too large by the amount of room taken up by the molecules, so we should expect the product pV to come out on the large side at high pressures on account of this.

The second assumption was that the molecules bounce off each other without any real interaction apart from the collision itself. This is not really a very sensible assumption since we have argued in Chapter 1 that solids and liquids are as they are *because* of forces between molecules. Presumably these forces still act when the molecules are in the gaseous state, though only to a marked extent when the molecules are forced close together. This will give another reason for real gases to fail the perfect equation when pressure is high and it is under these extreme conditions that the behaviour of real gases deviates most from the simple predictions of molecular theory.

These two effects and others of a more complex nature can be allowed for and very good agreement is then obtained between theory and experiment. The molecular theory stands the test of comparison with real measurements and the wonder of it is not that the simple results are slightly imperfect, but that they are almost right first time!

4.10 A major problem – how to measure temperature

We now return to a problem that has been hinted at earlier when it was necessary to have a temperature measuring device in the gas experiments, Sections 4.3 and 4.4. Having decided that temperature is related to the average molecular kinetic energy, we still have the problem of how to measure it.

Whenever we set up a method for measuring a quantity in physics we have to decide first on a standard size for it. A metre means a certain number of wavelengths of a definite coloured light. A second is the time for a definite number of vibrations of an atomic clock. A kilogram is a prototype lump of alloy kept in Paris. Similarly an amp, a volt, an ohm, a newton can all be fixed in relation to each other and we know what we mean by a unit amount of any of them.

But this is not yet sufficient. We need to be able to subdivide or extend the range beyond a one-sized chunk; what does $\frac{1}{2}$ a metre mean? How do we fix the position of the $\frac{1}{2}$ mark? With distance there does not appear to be much trouble, yet what we are tacitly saying is that two sticks $\frac{1}{2}$ metre long can be fitted end-to-end to form a metre, an assumption at least in accordance with our senses. This though is the whole point – we ultimately rely on our senses to judge multiples and sub-multiples of any basic unit. In this connection temperature poses the problem in an acute form.

To decide on a standard temperature interval is not hard. Ice-water has been called 0°C for a long time and the steam from boiling water under normal conditions has been called 100°C. This is the basic difference in hotness – 100 grades between the ice and steam points. But what does 50°C mean? How will we judge halfway in hotness?

We must find something whose measured value alters with hotness. There are literally thousands of alternatives to choose. Here are some:

> the length of a mercury thread in a narrow tube;
> the pressure of a trapped volume of air;
> the resistance of a platinum wire;
> the emf of a thermocouple;
> the vapour pressure of liquid helium;
> the light from a furnace;
> the surface tension of a liquid.

Suppose we measure one such quantity, say the resistance of a piece of platinum wire, and it has the value 30 ohms (Ω) in melting ice and 40 Ω in boiling water. We can then say that a temperature of 50°C will mean that degree of hotness at which the wire has a resistance of 35 Ω. We are using the graph of Fig. 4.12 really, from which the temperature can be read off if we know the resistance value. The two points at 0 and 100 are sufficient to fix the line of the graph. We are in fact measuring the temperature by assuming the chosen property, in this case the resistance of the platinum wire, varies uniformly with hotness. This seems to be a perfectly reasonable assumption.

However, when a different quantity is measured, say the length of a mercury thread in a narrow tube, we would want to say that 50°C means that hotness which gives a length halfway between the lengths

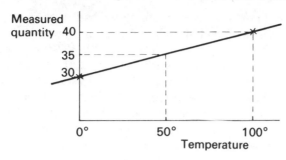

Fig. 4.12

at 0 and 100°C. This uses exactly the same procedure as before, but the two 50° temperatures would not necessarily be equally hot!

To understand why, think of a mythical problem of marking a ruler. We have a set of equal weights which we can use to stretch a spring or elastic band. The spring's length could be used to mark the ruler as follows. The ruler is placed alongside the weight-hanger – Fig. 4.13 – and the top level to be called 0 marked. An additional 10 weights are placed on the hanger and the position now is marked 10 – Fig. 4.14. What does position 5 mean? Clearly the place to which 5 extra weights will stretch the spring – Fig. 4.15.

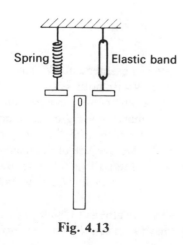

Fig. 4.13

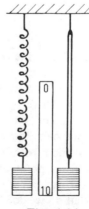

Fig. 4.14

Suppose instead of using a spring we had used an elastic band. Imagine one which will stretch as far as the spring under 10 extra weights, giving the same marks for 0 and 10. Will 5 extra weights on the elastic band take it to the same place as 5 extra weights did for the spring? Not necessarily. It depends on the spring and the elastic band – Fig. 4.16. Each one will give a perfectly valid way of marking the ruler, though they might only agree with each other at 0 and 10. Length measured by assuming a spring extends uniformly with load will not necessarily coincide with length measured by assuming an elastic band stretches evenly with load.

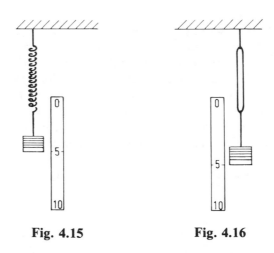

Fig. 4.15   **Fig. 4.16**

This is just the problem in temperature measurement. We have to use some quantity, *assume* it changes evenly with hotness and make a scale of temperature on which halfway means half-measured-value-hotness. There is no reason why 50°C marked using a resistance thermometer should agree with 50°C marked using a mercury thermometer (the sign °C merely fixes the average *size* of the degree, 100 of them between the ice and steam points): the actual size may vary from point to point according to the way in which we are measuring temperature.

What do we do? Nearly all gases behave in almost the same way as far as pressure and volume go, as we have seen in early sections, so a

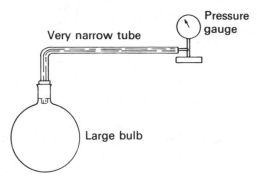

Fig. 4.17

gas may well be the best thing to use as our temperature-measuring quantity. Hydrogen most closely follows the simple theory, so this is the most suitable gas. But it does still appear to be an arbitrary choice; we could use the resistance of a platinum wire equally well. Yes, true enough, but in higher reaches of physics there are things called 'heat engines' which can also be used to fix a scale of temperature. It turns out that the theoretical heat engine scale is identical with the *perfect* gas scale to which all real gases more or less approximate. This provides an added reason for using a gas thermometer rather than a resistance thermometer for our standard device.

Fig. 4.17 shows a simple gas thermometer. The scale on the face of the pressure gauge can be marked directly in temperatures rather than pressures, and if we are fussy it can even be corrected for the fact that any real gas we put in the bulb will fall short of being perfect for reasons suggested in the last section. A thermometer like this, but of more sophisticated design, is in fact our standard hotness measurer and other thermometers ought, for accurate work, to be marked by direct comparison with one.

5

Waves and What They Do

5.1 First ideas

If you ask a friend to describe a spiral staircase he will invariably illustrate his answer by some circular upward movement of his hand. It is very difficult to give a clear picture of a spiral staircase without using gestures to help visualise it. In the same kind of way a person asked to describe a wave will most likely wiggle his hand vaguely up and down and hope the idea is clear.

The most common encounter with waves that all of us have is in the bath. Water waves are very familiar, yet we glibly accept the terms sound waves, light waves and radio waves, and we even have to cope with radar waves on the highway. These all have many features in common, as we shall see.

First and foremost, waves *travel*. This means they have a **velocity**. Light in free space zooms along at 300 million metres per second. Sound in normal air travels at roughly 340 metres per second. Water waves can travel at a variety of speeds but generally about 0·2 metres per second. The exact velocity for any sort of wave depends on the material it is going through, or the local conditions. Light travels more slowly in water than in air; sound travels more rapidly in water than in air.

Secondly, there is something about waves which occurs *regularly*. Crest follows crest, bump follows bump, and continues like this until either the energy of the wave is changed to other energy forms or the source of waves is switched off. This would also apply to the output from a machine gun, of course, and for a few lines further we will follow this similarity. Waves follow each other regularly, often a definite *time* and often a definite *distance* apart, just as the bullets would be chasing each other maybe $\frac{1}{10}$ second later and 2 metres apart. These characteristic numbers give the idea of **frequency** – 10 bullets per second, and of **wavelength** – 2 metres between bullets. Applied to a wave, frequency means the number of waves which pass a given point in 1 second, and is usually expressed in *hertz* which simply means 'per second'. Wavelength means the shortest distance between similarly moving points of a wave. The bullet figures give us some more information. If they are 2 metres apart and succeed each other after $\frac{1}{10}$ of a second, how fast must they be going? 2 metres in $\frac{1}{10}$ second means $20 \ \mathrm{m \ s^{-1}}$. This 20 is 2×10 and the 10 is the number of bullets per second, i.e. their frequency. So we have:

$$\text{Frequency} \times \text{Wavelength} = \text{Velocity}$$

$$\boxed{f\lambda = V}$$

This result is necessarily true for *all* kinds of waves because of the way we define frequency f and wavelength λ.

Let us use it on radio and sound waves with which we are more familiar. Radio waves go at the speed of light, in other words, $V = 300\,000\,000 \ \mathrm{m \ s^{-1}}$; a BBC programme goes out on a wavelength of 1500 m on the long wave. So we can find the frequency f in hertz:

$$f \times 1500 = 300\,000\,000$$
$$f = \frac{300\,000\,000}{1500} \ \mathrm{Hz}$$

$$= \underline{200\,000 \ \mathrm{Hz}}$$

This means that these waves, roughly a mile long, whizz past at a rate of 200 000 every second!

The numbers on an FM radio using the VHF waveband are frequency numbers, around 90 million hertz or 90 MHz. The velocity is the same, of course, so the wavelength can be found this time:

$$90\,000\,000 \ \lambda = 300\,000\,000$$
$$\lambda = \frac{300}{90} \ \text{metres}$$

$$= \underline{3 \cdot 3 \ \mathrm{m}}$$

The waves are much shorter this time and this point will explain later (Section 5.8) why only one transmitter is needed for the whole country on 1500 m while VHF transmitters are dotted up and down the land, usually on high ground, broadcasting on about 3 metres.

The frequency used by a man's normal speaking voice is around 150 Hz, so with a velocity of 340 m s^{-1} this gives a wavelength λ of:

$$\lambda = \frac{340}{150} \text{ metres}$$

$$= \underline{2\cdot3 \text{ metres,}} \text{ nearly.}$$

5.2 What are waves?

This is the hard thing to explain. The following sections will deal with what waves *do*, but we should try to put the idea of a wave into words and pictures first.

If you persuade a friend to shake the end of a long rope regularly and you take a flash photograph of the result, you will obtain the familiar wavy pattern (Fig. 5.1). This is a picture of the wave at one particular instant when the photograph was taken. It shows that, instead of the rope lying straight out in the undisturbed position, different parts of it have been moved up and down, one way or the other. Of course the distance AB is the wavelength λ. So is PQ, and any other length measured between corresponding points on the wave.

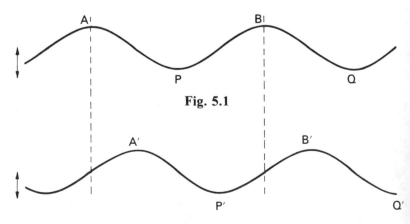

Fig. 5.1

Fig. 5.2

Suppose another picture is taken 2 seconds later. The wave will have moved on now to the position shown in Fig. 5.2, but the wavelength *AB* or *PQ* will still be the same. The odd thing about waves is that, although they progress from left to right along the rope, the rope itself does not travel with them! The moving thing which we call a wave is merely the *shape* of the rope which has altered so that the humps and hollows are in different places. What has happened to the rope? If we superimpose the two pictures (Fig. 5.3) we see that each part of the rope has moved up or down to its new position, perhaps up *and* down, but certainly not sideways. So we have the odd situation where each bit of the rope goes up and down while the wave travels sideways.

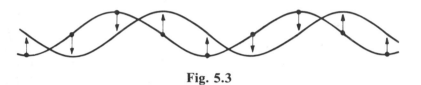

Fig. 5.3

5.3 Transverse waves

This is the commonest kind of wave which is easiest to understand and turns up in rope waves, water waves and in light waves, though in this last case you cannot take a photograph and get anything like Fig. 5.1 or Fig. 5.2. The kind of wave like this, where the thing that carries the wave vibrates at right angles to the way the wave travels, is called **transverse**. These are the familiar kind; they look wavy and all the properties of velocity, wavelength and frequency have clear and obvious meanings.

5.4 Longitudinal waves

There is another kind of wave, which occurs in sound, a kind of pressure wave. Instead of a gas having a constant pressure at all points in it, due to some regular disturbance there are points where the pressure is high or low. The pressure is raised by the gas's being temporarily compressed, and lowered by its being slightly rarefied. The compressions and rarefactions are caused by the source of the disturbance – a loudspeaker or the human vocal chords or the reed of a clarinet or the lips of a trumpet player.

Imagine some membrane flapping backwards and forwards at point A in Fig. 5.4. This diagram shows what would happen to a chain of particles connected together by loose springs if the end one were made to follow the movements of the membrane. The successive lines show the particle positions at regular moments of time and progressively the disturbance is passed down the chain, rather like the clanking noise of trucks bumping together under the pushes of a shunting engine.

The trouble with the picture at line 10, say, is that it does not look anything like a wave; but here is the moral of this section – any kind of disturbance which occurs at regular intervals and which progresses through space and time can be called a wave. The frequency means the

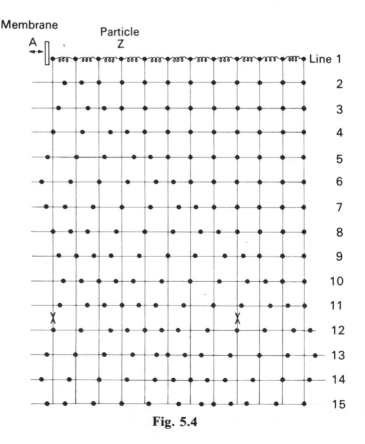

Fig. 5.4

number of disturbances of the same kind passing every second; the wavelength is the distance between consecutive corresponding points in the disturbance; and the velocity is the distance travelled by one point in the disturbance in one second. In the pressure wave drawn here the wavelength would be XX.

By following the career of one of the particles in the diagram, say particle Z, as time goes on it is seen to vibrate backwards and forwards, and the wave has reached to the end of the line. If the wave carrier vibrates along the same direction in which the disturbance travels, we have a **longitudinal** wave. Sound is an example of a longitudinal wave motion.

Another way of looking at a situation such as line 10 is in terms of pressure. Considering the spacing of the particles along the row and remembering that they began equally spaced, there are regions where the particles are more closely packed than normal and the springs between these would be compressed, and other regions where the gaps between particles are larger than normal whose springs would be stretched. This is drawn again in Fig. 5.5 and is the nearest thing we could get by photography to a real picture of this type of wave.

If a graph is drawn showing how the pressure varies along the line we might have a curve like Fig. 5.6, which is at least wavy and makes us feel happier about calling the row of dots a wave.

Fig. 5.5

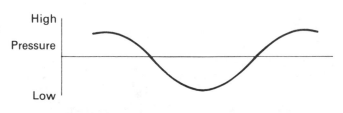

Fig. 5.6

What do waves do?

Nearly all the general properties of waves can be illustrated using water waves in a ripple tank. The next few sections will describe experiments with water waves, but the behaviour will be general to all types of waves, and examples from other types will be mentioned to illustrate the point.

A ripple tank is a rectangular or circular container for water, quite shallow and with a transparent bottom. Some means of making waves is needed and most commonly this is done by a boom carrying a small electric motor with an off-centre load. The speed of the motor can be regulated and the ripples are projected on to a white surface by using a small electric lamp over the tank or, alternatively, on to the ceiling if the lamp is under the tank. Fig. 5.7 is an example of a manufactured type, but they are not difficult to make out of simple equipment and materials.

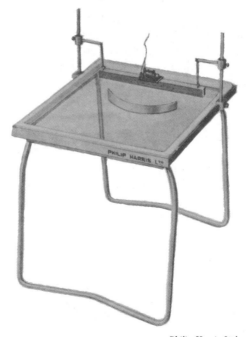

Philip Harris Ltd

Fig. 5.7

5.5 Waves go forwards

When the tank is ready and level and the water surface still, a 'pulse' can be sent outwards across the water by lightly touching the surface with a pencil or a finger. The ripple pattern, at close intervals, would be like the set of diagrams of Fig. 5.8. (A good tank will not allow much reflection from the sides.) As time goes on the ripple goes further out from the point where it was started at a steady speed, but decreases in 'strength'. This is what happens when a gun is fired, the sound of it travels outwards in all directions but gets fainter as it goes further away.

If a single dipper is fitted to the boom and the height of it adjusted so that the dipper is just in the water we get a stream of circular ripples when the motor is switched on, as in Fig. 5.9. This is similar to a lamp being switched on and left on, or a radio or television transmitting station operating continuously.

If the boom itself is lowered until it just touches the water surface all along its length and the motor switched on again, the pattern of ripples looks like Fig. 5.10. They are straight, equally spaced and travelling along the tank. If the motor is speeded up the picture changes to Fig. 5.11. The waves still travel at the same speed (this is decided principally by the depth of the water – see also Section 5.13), but the frequency has been increased by the faster motor speed. This will have to give a reduced wavelength since the product frequency × wavelength

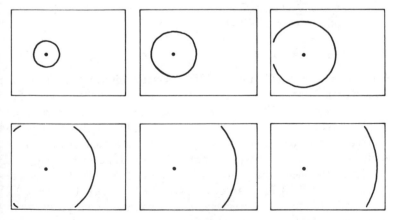

Fig. 5.8

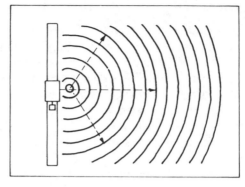

Fig. 5.9

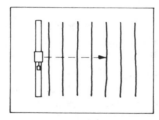 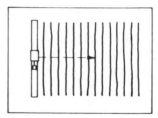

Fig. 5.10   **Fig. 5.11**

is still the same. The ripples are closer together. High pitch sounds have a shorter wavelength than low ones.

5.6 Waves bounce off barriers

If the boom is touched once while in the water without the motor on, we can get a straight pulse to travel down the tank and its progress can easily be followed. With a straight barrier placed across the tank the ripple bounces off like Fig. 5.12. As we expect, the angles coming into and receding from the barrier are equal. If the motor is switched on the pattern is more complex with two sets of ripples, one approaching the barrier and one leaving it, and a honeycomb effect is produced where they cross – Fig 5.13. In Chapter 6 we shall find that light bounces off a plane mirror in just the same way, so does sound when it produces an echo from a building.

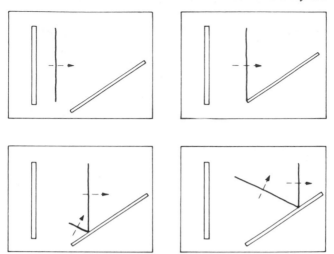

Fig. 5.12

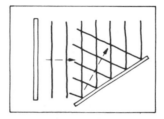

Fig. 5.13

When a circular pulse is used instead of a straight one each part of the ripple bounces off the barrier at the same angle as *it* arrives and a circular ripple leaves the barrier, but the centre is somewhere behind it – Fig. 5.14. The position of the centre of the reflected ripple is clearer if the barrier is put straight across the tank – Fig. 5.15. The centre is as far behind the barrier as the original centre of the ripple was in front. When you stand 3 metres in front of a mirror for dressing you are looking at a picture of yourself which is 3 metres behind the mirror. The moon's reflection in a lake is some 400 000 km under the water!

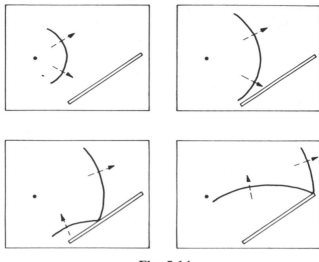

Fig. 5.14

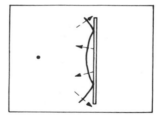

Fig. 5.15

5.7 Waves can be focused by curved mirrors

If the barrier is curved the effects will be different since the equal angle rule will apply at every point of the reflector. A straight ripple becomes bent by a curved reflector and not necessarily into a circle. It needs a specially shaped mirror to produce a circular reflection – Fig. 5.16. If the reflector is parabolic an incoming straight ripple is brought to a focus because the reflected ripple is perfectly circular (a circular barrier does not do quite as well, as it does not give a perfectly round reflected wave). The focusing effect is used by ornithologists to record a faint

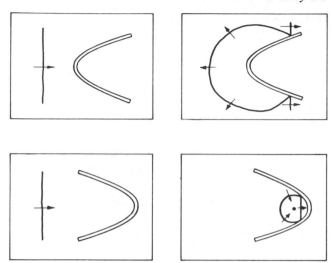

Fig. 5.16

bird song, and the bowl of the Jodrell Bank radio telescope has this shape to collect the faint radio signals from space.

If the process is reversed and a circular pulse started at the focus of one of these parabolic reflectors a straight ripple emerges – Fig. 5.17. Torches, spotlamps, car headlamps, radiant fire reflectors and many other simple devices use this arrangement.

The point about a specially shaped mirror is an important one when we consider the design of telescopes in Section 6.24.

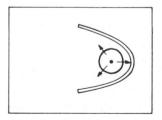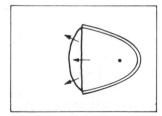

Fig. 5.17

5.8 Waves can bend round obstacles

This is probably the first new idea about waves and one which at first sight might not be thought a general property of all wave forms. In fact it is a trick which all kinds of waves do, but the magnitude of the effect differs so greatly that in many cases it simply is not seen.

With a ripple tank the way to investigate the effect is to set the boom vibrating with the motor so that a good set of straight ripples fall squarely on to some obstacle small enough not to block out all the width of the wave. The interesting result – Fig. 5.18 – is that ripples are distinctly seen in what ought to be the shadow area and at the far side of the tank the ripple may appear almost complete, though somewhat weakened. The wavelength does not change of course.

Using a bigger or smaller obstacle and altering the wavelength of the ripples by changing the motor speed produce some important changes in the pattern – Figs. 5.19 and 5.20. The crucial quantity which decides how much bending occurs is the size of the obstacle compared with the wavelength of the waves. Long waves bend easily around

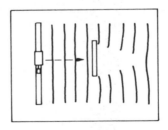

Fig. 5.18

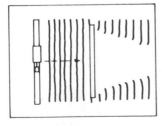

Fig. 5.19

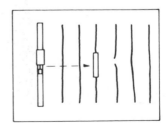

Fig. 5.20

small obstacles, short waves are hardly able to bend at all around large obstacles. This effect is well illustrated in the case of radio waves. To waves 1500 metres long most buildings and hills do not present much of a barrier, but the VHF transmissions on about 3 metres cannot bend easily around such obstacles. There is only one transmitter for the whole country on the long waveband whereas we need many VHF stations serving much smaller areas.

When we say that light travels in straight lines it fits in with our inability to see round corners. The wavelength of light is about $\frac{6}{10\,000}$ mm, so ordinary objects will provide for a negligible amount of bending. When light is shone at a very small object though, like a hair, there is definitely some light reaching the shadow area showing that there has been some bending – Fig. 5.21. A similar thing can be seen when light from a distant street lamp is viewed through the mesh of an umbrella cover – Fig. 5.22. The light bends, or is *diffracted*, round the thin fibres of the fabric.

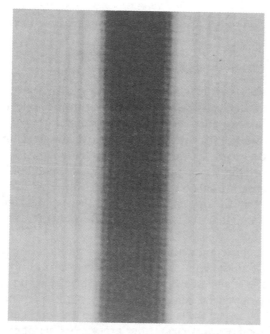

Cambridge University Press

Fig. 5.21 The 'shadow' of a very thin wire

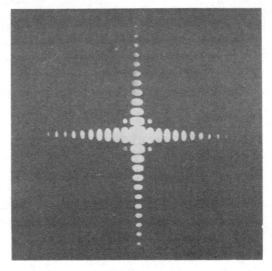

G. R. Noakes and B. K. Harris, Macmillan

Fig. 5.22   View of a distant lamp through a fine-mesh fabric or
gauze

5.9 Waves can go through small holes

Instead of using a barrier which blocks out a small part of the waves we
can send them through a gap which allows through only a portion of
the waves. We can alter the same dimensions in this case too, and
Fig. 5.23 shows how the patterns look for large and small holes and for
long and short waves. Again, the important thing is the size of the hole
compared with the wavelength of the waves. A small hole lets hardly
any long waves through, but short waves easily pass through a large
hole. Notice that the ripples do bend into the shadow areas as in the
last Section and that the wavelengths are not altered by passage
through the holes.

The main connection with other sorts of waves concerns light and
telescopes. Of all the light that a star sends to earth only that which
enters the tube of a telescope will enable it to be seen and the edges of
the light waves will be bent, even if only slightly. From the ripple
pictures it is clear that the larger the hole the less this bending is, so to
get a really good picture of a star the largest possible 'aperture' for the
telescope is required. In fact the dimension used to describe the
telescope is the size of the aperture or diameter. The '200-inch'

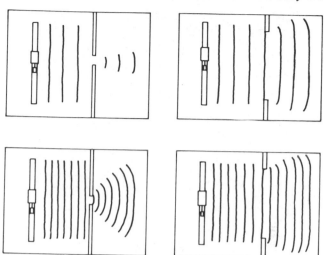

Fig. 5.23

telescope at Mount Palomar for instance gathers light over an area 200 inches across. The need for a large aperture for a telescope is dealt with again in Section 6.24.

The habit of waves to bend round obstacles or through small holes is called **diffraction** and is one of the trademarks of a wave. Sound obviously diffracts easily round ordinary objects, since we can hear round corners, and referring back to Section 5.1 shows that its wavelength is just about right for this to happen.

5.10 Waves can knock each other out

This happens when two lots of waves overlap and the proper word for it is **interference**. In one sense it is a bad choice of word because once the waves have passed the region where they overlap they continue on their way as if nothing has changed – Fig. 5.24.

For interference effects to be seen easily the two sets of waves have to be related to each other. In practice the best way to ensure this is to derive them both from the same source of waves. In a ripple tank we can fit two dippers side by side on the boom and both will be vibrated by the motor. Each dipper gives a set of circular waves, but the interest lies where they overlap and the picture looks like Fig. 5.25.

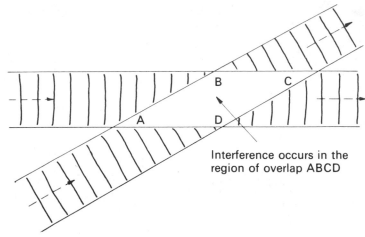

Interference occurs in the
region of overlap ABCD

Fig. 5.24

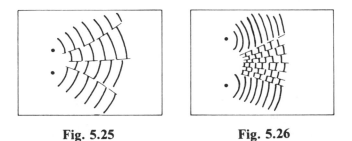

Fig. 5.25   **Fig. 5.26**

The feature of this pattern is that, although both sets of ripples are travelling across the tank, in the region where they meet there are distinct 'lines' where the water is undisturbed and remains undisturbed. Making the wavelength smaller or spacing the dippers further apart causes these doldrum lines to lie closer together – Fig. 5.26. They are not straight (actually they are hyperbolic) and they exist at intervals where the waves overlap. Between the quiescent lines the water is more than usually disturbed. It seems that the price for getting some patches of quiet water is that other areas are overworked. Really what has happened is that the energy in the two sets of waves has been redistributed in the interference zone, not wiped out altogether.

Can this odd effect occur with other types of waves? Certainly with radio and sound waves, and even with light waves if care is taken. Some radio waves reach us by two routes, one along the surface of the earth and one by reflection high up in the atmosphere – Fig. 5.27. If the receiver happens to be at one of the quiet zones for these two sets of waves reception is very bad. More commonly reception fades in and out from certain European stations as the reflecting layer in the atmosphere shifts about slightly.

In sound the effect is easily heard by setting up two loud-speakers connected together a few metres apart and driving them from the same amplifier or oscillator. (This is best done through an open window as reflections from the walls and ceiling of a room tend to obscure the effects.) If you walk across the pattern a few metres from the loudspeakers you walk through zones of silence and zones of loudness – Fig. 5.28.

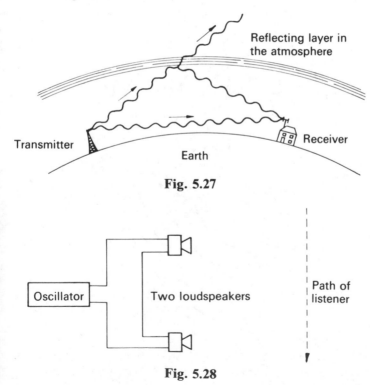

Fig. 5.27

Fig. 5.28

5.11 How is the interference pattern formed?

A closer look at the ripple pattern is needed now to see how it is formed: Fig. 5.29 shows the details of it. The solid lines represent the crests of the water ripples and the broken lines represent the hollows. Alternate crests and hollows are made by each dipper and the points marked with a cross are where a crest from one dipper meets a hollow from the other, in other words where a solid line meets a broken line.

The crosses will be neither crests nor hollows, so the water there will

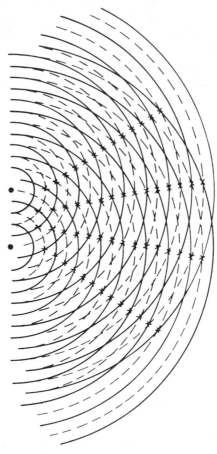

Fig. 5.29

be undisturbed along the lines of crosses. Where crest meets crest there will be a bigger crest and where hollow and hollow coincide a deeper hollow, so the water in these places will carry a larger disturbance than normal.

How far apart are the zero lines? One thing which decides this is the distance away from the two sources of waves as the lines clearly get further apart. The other two factors concerned are the wavelength of the ripples and the separation of the dippers. The connection between these quantities is:

$$\frac{\text{Distance between zero lines}}{\text{Distance from the sources}} = \frac{\text{Wavelength of the waves}}{\text{Separation of the sources}}$$

or in symbols (Fig. 5.30):

$$\frac{w}{D} = \frac{\lambda}{s}$$

(The proof of this relation need not concern us here. It is a piece of not very exciting geometry.)

For example, using dippers 5 cm apart generating waves 2 cm long we would have to go 25 cm from the dippers to get zero points spaced 10 cm apart.

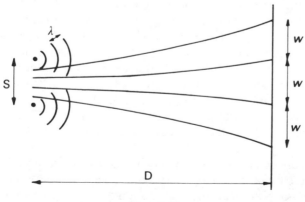

Fig. 5.30

5.12 Can light + light = darkness?

If the interference experiment is done with light there ought to be lines of darkness in the pattern! Two lights shining on the same area ought to give darkness in some places and extra brightness in others.

There are two difficulties in doing this. One arises from the result of the last section. For light, the wavelength $\lambda = 6 \times 10^{-4}$ mm and we might hope to get two lamps very close together, say $s = 1$ mm. Suppose we look 1 metre from this double-lamp arrangement, how far apart would the dark lines be?

$$\frac{w}{1000} = \frac{6 \times 10^{-4}}{1}$$

giving $\qquad w = \underline{0{\cdot}6 \text{ mm}}$

The lines will be very close together and easily missed. Nothing can be done to improve things by altering λ and to go much further away would reduce the brightness too much, so we have to try to place the lamps even closer together if we want a wider separation between the interference lines.

The other difficulty lies in the fact that the two sources of waves must be related to each other in some way and in the ripple tank this was arranged by having them both driven by the same motor. The way out of both these snags for light is to shine light through two slits which are very close together using only one lamp. Thomas Young was the first man to do this in 1802.

A piece of glass painted black will do with two fine lines scratched in the paint very close together using a pin and a ruler – Fig. 5.31. The lamp can be a car headlamp, the kind with a straight filament, and the only tricky part to the experiment is to get the filament and the twin scratches parallel to each other. A darkened room and some barriers to cut out stray light from the lamp are needed and a little patience will produce a pattern on the screen like Fig. 5.32. A coloured filter will also help to make the pattern more distinct though fainter.

Another way to see these stripes is to hold the glass slide very close to the eye and look at a distant lamp – Fig. 5.33.

There are definitely more than two bright patches and if a card is very carefully manoeuvred in sideways to cover only one of the slits the pattern disappears, showing that it is a phenomenon associated with both slits.

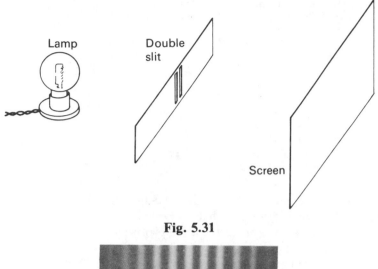

Fig. 5.31

Cambridge University Press

Fig. 5.32   Double slit interference pattern

This was a crucial experiment in convincing scientists that light really does behave like a wave because it is very hard to imagine anything else which can cancel itself out in some places and reinforce itself in others.

5.13 Waves may alter their direction if they alter their speed

To do this successfully with water waves requires a careful control of the depth of water. A glass or plastic sheet must be put in the tank and

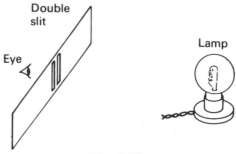

Fig. 5.33

the water added until it will only just cover the sheet. A side view of the tank is shown in Fig. 5.34. With the boom making straight ripples and the glass sheet square on to them the picture is like Fig. 5.35.

The waves are closer together over the glass sheet where the water is shallower, in other words the wavelength is reduced. What about the frequency? This is decided by the speed of the motor driving the boom and is nothing to do with the water; it still makes the same number of

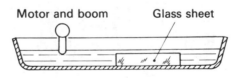

Fig. 5.34

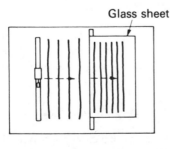

Fig. 5.35

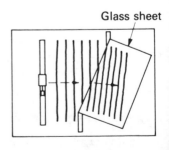

Fig. 5.36

ripples every second, so the frequency must be the same as before. We know that the frequency × wavelength = velocity (Section 5.1), so if frequency remains the same and wavelength goes down, the velocity must also go down. The waves get closer together *because* they slow down, and they slow down because the water is shallower.

So far there is no change of direction, but this time we set the glass sheet so that its edge is oblique to the ripples. Now one end of the ripples reaches the shallower region before the other end. What happens? Fig. 5.36 shows the result. (It is a good idea to block out the corners of the tank to avoid confusing patterns.) The waves change their direction; they slew round on reaching the part where they go slower. Of course they slew the other way when they enter the deeper part again.

This sort of bending is common with light. When a beam of light is shone into a transparent substance like water or glass or clear plastic it bends in just the same kind of way – Fig. 5.37. The inference from this is that light travels more slowly in glass than air in order to bend as it does.

Bending of waves due to a change in velocity is called **refraction** and in light it is used in lenses and prisms to make optical instruments – spectacle lenses, microscopes, binoculars, etc. These are dealt with in Chapter 6 in detail.

Sound waves can be refracted too. Sometimes it is very easy to hear sounds coming from a long way away while nearby noises are harder to hear. This happens when warm air lies on top of cooler air – see Fig. 5.38. Sound travels more rapidly in the warmer air and the waves can bend all the way round and end up coming downwards again. The opposite will happen if the air gets cooler as you go up – Fig. 5.39.

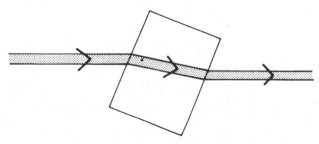

Fig. 5.37

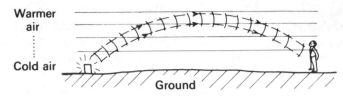

Fig. 5.38

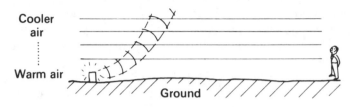

Fig. 5.39

5.14 Waves that stand still

A special kind of interference happens if the two sets of waves are similar in wavelength but travelling in opposite directions. With a ripple tank this could be done by having two booms each driven by motors at the same rate, but it would be very hard to get the motors in step. It is much easier to use one set of ripples and their reflection from a straight barrier. This time the pattern contains lines only half the wavelength of the original waves apart, but the main thing about them is that they do not move either backwards or forwards – they are called **stationary** or **standing** waves. Fig. 5.40 shows the pattern.

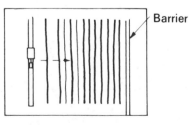

Fig. 5.40

A series of drawings show how these patterns arise. Think of two waves similar to each other and travelling in opposite directions like *A* and *B* in Fig. 5.41. Both of these will make humps in the same places and hollows in the same places, resulting in higher humps and deeper hollows, curve C_1. If we wait until *A* and *B* have separated by half a wavelength, in other words wait $\frac{1}{4}$ of a period, and add them up again we get curve C_2 which is a straight line, no disturbance. The humps of *A* coincide with the hollows of *B* and give no wave at all. A $\frac{1}{4}$ period later we get the next position and the total at that time is curve C_3, a

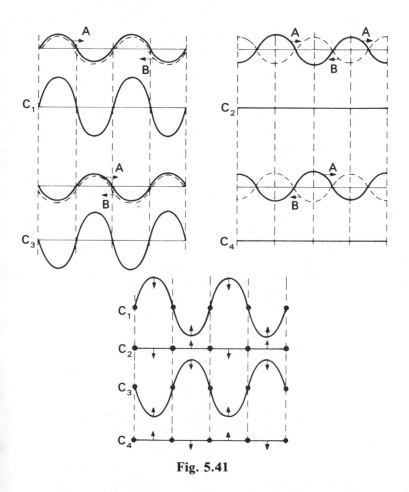

Fig. 5.41

big one but opposite to C_1. Next comes C_4 then C_5, etc. Looking at C_1, C_2, C_3, C_4, etc. in turn tells us that the wave is in 'loops' with big humps and hollows alternately spaced and points between them acting as 'pivots'. These are marked with a dot.

The up and down movement might take place so rapidly that all an observer sees is the set of 'pivot' points with a blur in between each one. In the ripple tank the undisturbed 'pivot lines' are clearly seen and looking at these diagrams they are spaced only half a wavelength apart.

A rope can easily be made to wiggle at slow motion in this loopy pattern. If you tie a reasonable length, 4 or 5 metres, to a door knob and wiggle it slowly you can make the rope go up and down in just one loop – Fig. 5.42. If you wiggle it faster you can make it vibrate in two, three, four or more loops, but some parts of the rope, the 'pivot points', always stand still and there is no appearance of waves travelling either way along the rope.

Stationary waves in sound can often be experienced while watching TV. Television sets make a very high pitch noise when operating and sometimes the viewer will notice that the note becomes louder and softer as he moves his head and he can make it almost disappear for certain positions of the head. What he is doing is to put his ears at the 'pivot points' of the standing waves caused by reflections of the note by the walls and ceiling of the room. In these places he hears a minimum

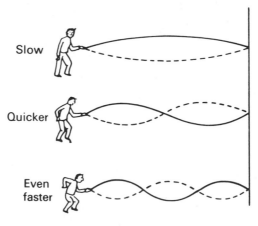

Slow

Quicker

Even faster

Fig. 5.42

of sound and often we position our heads like this without thinking about it.

5.15 Flat or 'polarised' waves

If you shake a rope which is tied to a post several metres away you will make transverse waves on it, the rope vibrates sideways or up and down and the waves travel along it. By indiscriminate shaking you get a general waviness going along the rope with no special direction about the way the rope vibrates. If you take care though you make the rope vibrate only vertically, or only horizontally, or in a diagonal direction and the wave looks 'flat'–Fig. 5.43. We call such a wave a **polarised wave**, and would say it is vertically polarised, or horizontally polarised. There is a definite direction of vibration about it, in this case imposed by the person doing the shaking.

The possibility of getting a polarised wave can only arise with transverse ones; longitudinal waves have their vibration to and fro along the direction of travel so cannot be polarised like this. There is

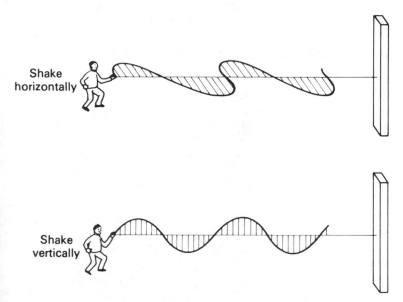

Fig. 5.43

only *one* possible direction for longitudinal waves to vibrate so polarisation does not arise for them.

With a very long rope, a good deal of patience and a couple of assistants the main features of polarised waves can be demonstrated. You need a few strips of wood, half-metre rulers are suitable or dowelling will do. The idea is to have two 'gates' for the rope waves to go through made of two pieces of wood held either side of the rope and quite close together – Fig. 5.44. The person at the end of the rope shakes it at random producing waves which vibrate in all directions. If the first gate is vertical, only the vertically polarised waves will get through, so the rope beyond the first gate vibrates only up and down. An interesting thing happens if a second gate is arranged for the waves to enter. If this second gate is also vertical the waves get through without hindrance, but if it is set horizontally the rope hardly moves beyond the second gate – Fig. 5.45. With the rulers set diagonally at the second gate a weak wave gets through polarised diagonally.

There are many substances which can do this to light waves, the commonest one nowadays is polaroid. If two pieces of polaroid are held up to the eye it is possible to turn them so that hardly any light gets through. Their polarising directions must then be at right angles like the vertical and horizontal gates for the rope – Fig. 5.46.

In fact there is quite a lot of polarised light about. A sunny blue sky sends some polarised light to earth, especially at right angles to the

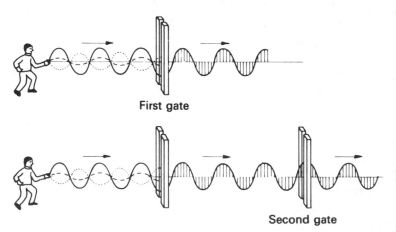

First gate

Second gate

Fig. 5.44

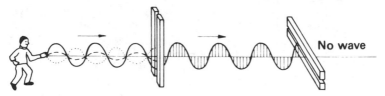

Fig. 5.45

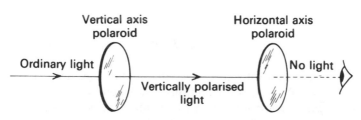

Fig. 5.46

sun. If you look at right angles to the way the sun is shining through a piece of polaroid and turn it, the blue becomes much darker when the polaroid direction is right – Fig. 5.47.

A lot of reflected light is polarised, sunlight on the sea or a pond for instance, or from a window or the pavement, and this is the idea behind polaroid sunglasses. They are set to eliminate the horizontally polarised light reflected from horizontal surfaces and so cut out dazzle – Fig. 5.48. (A strange idea here is to wear these sunglasses for night driving if it is wet; they will cut out the dazzling reflections from the wet surface of the road. In fact we could easily design a perfect

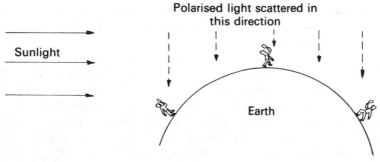

Fig. 5.47

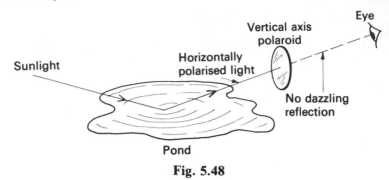

Fig. 5.48

anti-dazzle system for cars with polaroid fitted in headlamps and windscreens. Why do you think this is not done?)

We often see television aerials set in different directions in different parts of the country. This is because the different transmitters send out polarised waves, some using them vertically polarised, others horizontally. There is no danger of overlapping or interference between two adjacent stations if they are oppositely polarised.

Resonance and musical instruments

We will spend a few pages looking at one special application of vibrations and waves, namely the basic ideas behind a few musical instruments and why most of them sound distinctive.

5.16 Resonance

This idea is a very important one in physics and can be used in many clever ways, or it can be a nuisance and require elaborate precautions to avoid its happening.

If you blow across the mouth of a bottle you can get it to make a note which lasts as long as you blow – Fig. 5.49. The frequency of the note depends on the volume of the bottle and the dimensions of the neck. The air inside the bottle has its own 'natural' frequency. Another way to get the bottle vibrating is to tap it lightly with a hammer. Young children often make a model xylophone using bottles partly filled with water, each bottle has its own natural frequency and makes its own sound – Fig. 5.50.

Fig. 5.49

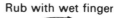
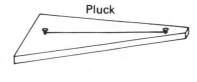

Fig. 5.50

If you pluck a tight string or wire it vibrates visibly and makes a note depending on its length, tightness and thickness. For a given string, you get a definite note, its natural way of vibrating. You can make a wine glass 'sing' by rubbing round its rim with a wet finger – it gives out its own natural frequency – Fig. 5.51 and Fig. 5.52.

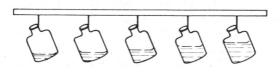

Fig. 5.51　　　　　　　　　**Fig. 5.52**

Benjamino Gigli was supposed to have shattered a wine glass by singing the right note, perhaps an apocryphal story, but I personally have shattered a glass lamp shade (not on purpose!) by playing a certain note on a clarinet – Fig. 5.53. Often when travelling in a car things left on the parcels shelf start to rattle at certain engine speeds, and sometimes in an inexpensive radio or record player every time a certain note is sung or played it sounds fuzzy or distorted. A few men can shift a car from an awkward spot even if it is locked, by bouncing it up and down and pulling sideways when the car is 'up'.

All these interesting events are examples of **resonance** where two frequencies just happen to fit – the one at which the wine glass or lamp shade or bunch of keys or car would naturally vibrate, and the one at which it is forced to vibrate by some other agency. When these two coincide we have resonance; the two things concerned resonate together and we can get the vibrations building up perhaps to catastrophic dimensions. One famous example of that was the spectacular collapse of the Tacoma Narrows suspension bridge in America under the progressive effects of high winds.

It is possible to get electrical circuits to vibrate naturally (using a capacitor and an inductor) and if the circuit is linked to an aerial picking up radio waves the electrical vibrations will only be large for that station whose frequency coincides with the natural frequency of the circuit. This is how it is possible to tune-in to only one radio

Fig. 5.53

station, although lots of stations are broadcasting at the same time though on different frequencies.

On a simpler level if you push someone on a swing you will only get them swinging high if you time your pushes to coincide with the swing's motion – you and the swing must resonate to give a large amplitude of swing. A diver can get quite a lift off a diving board if he jerks himself in step with the board.

Resonance can sometimes be a nuisance when using a record player or hi-fi set. The loudspeaker cabinet must be designed so that it does not have a natural frequency of vibration anywhere within the range of notes it will have to reproduce because, if it does, those notes will be much louder than the rest.

5.17 How can we see a sound wave?

In the next few sections we will need to meet a device called an *oscilloscope* – Fig. 5.54. For the time being we shall treat it just as a box which will do a certain job for us, but later in Chapter 9 its workings will be explained.

This is rather like a small television set with a screen which will trace for us any *electrical* signal sent into the oscilloscope. To use it to look

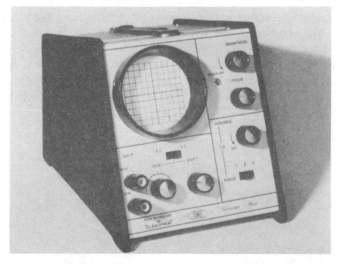

Telequipment Ltd

Fig. 5.54 Oscilloscope

at sound waves we must first change the sound vibrations into electrical vibrations, and a microphone is the thing to do this. When a sound is made into the microphone the oscilloscope draws for us the shape of the sound wave which enters the microphone (assuming it is a good microphone) – Fig. 5.55.

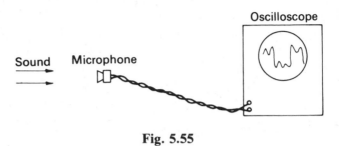

Fig. 5.55

5.18 Loudness

Using an oscilloscope and microphone as above the effect of different loudnesses is easily seen – a loud note gives a much taller wave pattern than a quiet note – Fig. 5.56. The maximum displacement of the pattern from the centre line is called its **amplitude**. The louder the note the bigger the amplitude of its wave, which means the air molecules are shifted further from their normal positions than for a quiet note. (Careful here: loudness and amplitude are not the same thing. A wave can have a big amplitude, but a deaf man will say its loudness is zero! And we are all of us deaf to a certain extent to certain notes.)

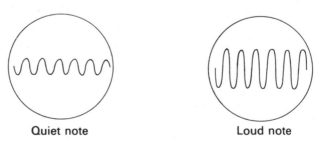

Quiet note   Loud note

Fig. 5.56

5.19 Pitch

This time we can sing or use an instrument like a tin whistle or flute to make a high note and see how its wave differs from that of a low note. The patterns show that more waves are drawn on the screen for a high note than for a low one – Fig. 5.57. The thing to remember about oscilloscope traces is that they are *time* graphs, so in this case it means that more waves happen in a certain time for high sounds than for low ones or, in other words, high notes have a higher frequency. The bigger the frequency the higher the pitch of the note. (Careful again – pitch and frequency are not the same thing. The idea of pitch depends on the listener and tone-deaf people find it hard to pitch a note. In music pitch and frequency are connected by the decision to say that two notes, one of whose frequency is double the other's, are an octave apart.)

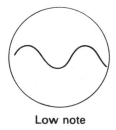
Low note

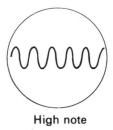
High note

Fig. 5.57

5.20 Quality or timbre

It is clear to most people that although a singer, a violin, a piano, a flute, a clarinet, a trumpet could all be playing the same note equally loudly they nevertheless would sound different. The quality of the note each instrument makes tells us what the instrument is. This is sometimes called the *timbre* of the note.

If a collection of instruments like these play in turn into a microphone and oscilloscope they produce distinctive waves which differ in their shape or profile; some are smooth and rounded, others contain sharp spikes or oddly shaped waves – Fig. 5.58.

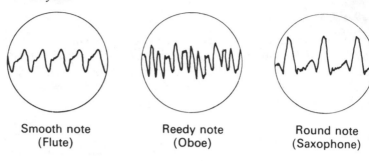

Smooth note   Reedy note   Round note
(Flute)   (Oboe)   (Saxophone)

Fig. 5.58

Why are there spikes and bumps on the waves from different musical instruments? The reason is that any vibrating object usually has more than one natural frequency of vibration. For instance a tight string is used to generate the sound in a violin, guitar, piano, zither, banjo, etc., and for a fixed length, tension and thickness the string can vibrate in a number of ways. If you take care you can make a guitar string, say, make a series of notes – you can touch it *lightly* at point X and pluck it at point Y for each of the positions drawn in Fig 5.59. Once it has been plucked the finger can be removed from X and the string will go on making that particular note until it dies down.

When the string is just plucked in the normal way or with a plectrum or bowed in the case of a violin the string vibrates in more than one of

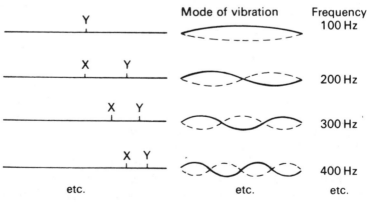

Fig. 5.59

these modes at the same time – if it were the first and third modes the string might vibrate as in Fig. 5.60 with the higher frequency called an **overtone** superimposed on the lowest one called the **fundamental**. In this case the overtone has three times the frequency of the fundamental and the resulting wave is what you get by adding these two waves together – Fig. 5.61. As you can see the shape of it has altered. Clearly by adding all the different overtones to the fundamental in their correct strengths you can get almost any shaped wave you wish. (Incidentally this is just what is done in electronic organs where the combinations of frequencies are chosen to try to imitate real instruments.)

Wind instruments too have their own set of overtones for a given length of pipe (which decides the fundamental frequency).

For a clarinet, for instance, the frequencies approximately follow a pattern like 100, 300, 500, 700 . . . , whereas for a trumpet they are like 100, 200, 300, 400 . . . These numbers are only approximate and their relative amplitudes differ from instrument to instrument; indeed, a player can often alter this factor as he goes along. The piano is a specially difficult instrument to imitate as its overtones do not fit a simple series and the note is struck percussively (with a hammer) and then dies away.

There is more to the question of how we recognise different quality musical sounds than we have discussed here. The majority of the

Fig. 5.60

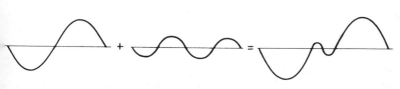

Fig. 5.61

information 'this is a clarinet' is contained in what are called the transient frequencies – these are the sounds made only while the instrument is beginning a note and they die away in a second or less, leaving the established pattern of overtones as displayed on an oscilloscope. Transients can be studied by making a tape recording at high speed, cutting off the bit of tape containing the beginning of a note and playing that back at a much lower speed. The 'starting off' noises can be heard more clearly then.

6

Light and Lenses

6.1 Does light go in straight lines?

If you ask most people what they know about the way light travels they will almost certainly say 'very fast and in straight lines'. How fast? $300\,000$ km s^{-1} very nearly. To the moon in 1·3 sec; to the sun in a little more than 8 minutes. Fast, certainly, by any standard.

Why do we say it goes in straight lines? Everyday experience and behaviour is based on the 'fact' that we cannot see round corners! We look 'straight' to an object we wish to scrutinise. We can see shafts of sunlight in the clouds, beams of light from torches or projectors in a smoky room, and the sun goes out during a total eclipse. These may sound very good reasons for light's straight line travel and we may even perform an 'experiment' like Fig. 6.1.

Three cards with small holes in them are lined up and moved so that we can see through all the holes. Without moving the cards it is then possible to push a 'straight' needle through the holes, so obviously the light was going 'straight' too! Well . . . maybe.

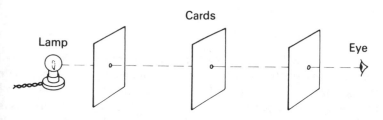

Fig. 6.1

The trouble lies in the word 'straight'. What do we mean by a straight line? How do we judge straightness? By lining it up to the eye and looking along it? That uses light though! By comparing it with a known straight edge? This is the chicken-and-the-egg situation! The point is that straightness means 'as light travels', so it is no wonder we can push a needle through the holes in the cards – imagine the difficulties if we could not!

Saying light travels in straight lines does not really achieve much. More to the point is that when viewing along a light beam we interpret it as having taken a straight path – even if it has not! This can be shown by carefully filling a long transparent tank half full of strong salt solution and half full of water with a little fluorescein to make the path visible – Fig. 6.2. The water is best put in first and the brine added slowly down a tube through the water to the bottom of the tank. Inevitably some mixing occurs where the liquids meet and a turbulent region is formed. A narrow beam of light aimed at this meeting layer can clearly be seen to wiggle about when viewed from the side of the tank. When it is viewed from the end of the tank though it appears to be a perfectly straight line!

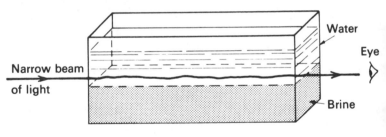

Fig. 6.2

6.2 A simple camera

What do we need to take a photograph? A light-tight box, a piece of light-sensitive photographic paper, a pin and plenty of patience. With a box some 15 cm long and a pin-hole in one end, a tolerable photograph can be taken with an exposure of some 5 or 10 minutes in fairly bright outdoor conditions. The photograph size depends only on the dimensions of the box, and all objects close to or far away are in focus at the same time without distortion at the edge of the picture. Fig. 6.3 is a photograph taken with such a camera using a small pinhole.

It is easy to see what happens – Fig. 6.4. Some light from each point of a bright object goes through the hole and affects one point only of the photographic paper. So dot-by-dot a picture is built up, rather like a newspaper photograph, where there is a one-to-one correspondence between points on the paper and points on the object. When this happens we say an *image* is made, and this can be processed in the normal photographic way with developer and fixer.

The snag with such a simple camera is that very little light enters

M. Riches

Fig. 6.3 Pin-hole camera photograph – small hole

Light-tight box

Small hole

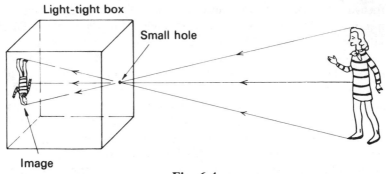

Image

Fig. 6.4

M. Riches

Fig. 6.5   Pin-hole camera photograph – large hole

through the small hole and a long exposure is needed, maybe several minutes. It obviously cannot be used to photograph moving objects. On the other hand, there are no focusing problems – all objects, no matter how near or far, are in focus together!

The solution to the intensity problem is to make the pinhole larger, but though this does give a brighter image (or shorter exposure time) it ruins the sharpness of the image. Each point on the object now sends light to a distinct patch on the back of the camera and all detail becomes blurred. Fig. 6.5 is the same photograph but with a larger pinhole. Fig. 6.6 explains why the blurring occurs, but it is still done by light going in straight lines.

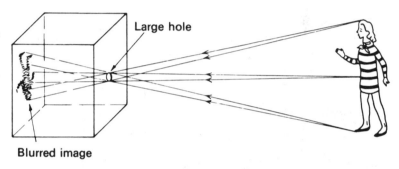

Fig. 6.6

6.3 Mirrors and reflections

Reasonably flat surfaces of almost any substance will often reflect sufficient light back for images to be seen 'in' them. The sun can be reflected in the sea, or polished tiles, or a plain window, or army boots! All we need is a high degree of smoothness and even if the bulk of a substance is transparent its surface can act as a mirror.

Some materials – notably metals – can be very efficient reflectors and send back 99 % of the light falling on them, and these are the ones normally used as mirrors. (Incidentally it may strike the reader as odd that the best *reflectors* of light are the best *conductors* of heat and electricity. This is not odd – all these processes involve the electrons of the metal and are all basically electrical properties.)

Confining our attention then to these good reflectors it is easy to

find out how light behaves when it meets one. A beam of light will
bounce off and always keep the angles coming in and going out
equal – Fig. 6.7. The posh way of saying this is that the angles of
incidence and reflection are equal and coplanar (i.e. the reflected ray
does not suddenly come up out of the paper if the incident ray lies
along the paper and the mirror at right angles to it).

The equal angle behaviour has several interesting consequences.
When you look at the reflection of your foot in a mirror on a wall, say,
your eye receives diverging light from every point on your foot. Fig. 6.8
shows the divergence exaggerated. There is no doubt the light path
suffers a sharp change of direction at the mirror, yet you think you see

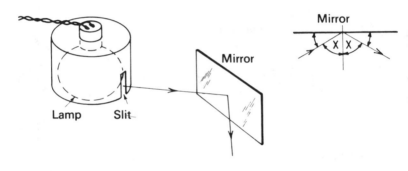

Fig. 6.7

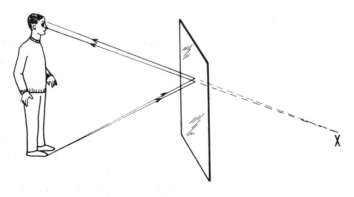

Fig. 6.8

an image of your foot where the diverging beam would have come from if it had been going in a straight line all the way – in other words at X. Where is X? Geometry or another experiment will help you decide. Two similar rods, one longer than the other, are arranged as drawn in Fig. 6.9. The shorter one is in front of the mirror, the longer one behind. What you have to do is to move the one behind so that no matter from what angle you view the reflection of the short rod it always appears as a continuation of the top of the long rod you can see over the top of the mirror. The reflection then must be in the same place as the long rod and direct measurement will show that both rods are equidistant from the mirror – in other words the image is as far behind the mirror as the object is in front. If you stand a metre in front of the mirror your reflection is two metres from you, one metre behind the mirror.

There is another curious thing about this reflection. The image is not really there. No light gets behind the mirror, yet that is where you see the image. This kind of image is called **virtual**, meaning *it appears to be* in a certain place, though no light really does come from there. The brain conjures the whole thing up because we order our lives according to 'light goes in straight lines', so that is where we see the image.

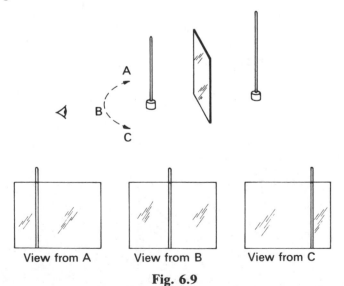

Fig. 6.9

Flat or plane mirrors have many uses as the reader will know, both functional and decorative. There is now even one left on the moon to reflect light back to a laboratory on earth. More mundane applications vary from dressing tables to kaleidoscopes to sextants to sensitive galvanometers to periscopes to driving mirrors to reflex cameras . . .

6.4 Curved mirrors

It is worth mentioning in passing that mirrors which are cut out of a spherical piece of glass or metal have certain advantages over flat mirrors for special purposes. Wing mirrors on cars are often convex ones – Fig. 6.10. They give a wider field of view than plane mirrors, though they do distort the reflected car's speed. Sometimes at awkward blind corners you will see a convex mirror so placed that a driver can see round the corner – Fig. 6.11.

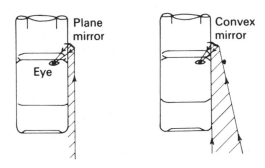

Fig. 6.10

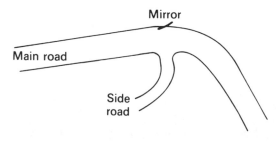

Fig. 6.11

Concave mirrors are used either for projecting a beam of light or for giving a magnified picture. Torches and car headlamps use a concave mirror to make a directional beam – Fig. 6.12. (For a decently parallel beam the mirror must be paraboloidal, not spherical – Section 5.7.) Shaving and vanity mirrors are slightly concave giving magnified pictures of the face. There is also a special use for curved mirrors in large telescopes – Section 6.24.

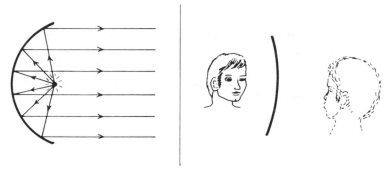

Fig. 6.12

In many ways the behaviour of concave and convex mirrors is similar to that of converging and diverging lenses respectively, which will be dealt with in detail later in this chapter.

6.5 Light bends on crossing the boundary between two substances

It is a common experience that swimming baths always seems less deep than they really are. A stick part-submerged in water appears to be bent where it enters the water. The reason for these effects is that, although light goes straight in one material, where it crosses from one material to another it suffers a sudden shift of direction.

A narrow beam of light can easily be shone at different pieces of transparent material to show which way it bends. In Fig. 6.13 the dotted lines are perpendicular to the surface where the light strikes the boundary, and going from air to glass the light bends closer to that line and coming out from glass to air it bends away from it. The dotted line is called the *normal* to the surface.

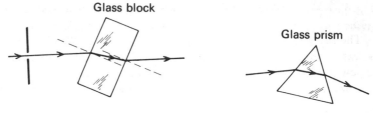

Fig. 6.13

Sometimes when trying to get out into air from glass or water the angle is too large and all the light can do is to bounce off the inner surface as if it were a mirror. This is called *total internal reflection* – Fig. 6.14. Coloured edges are often observed in the beams emerging from this *refraction* process, as it is called, especially when more than one bending occurs in the same direction, as in the prism of

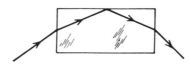

Fig. 6.14

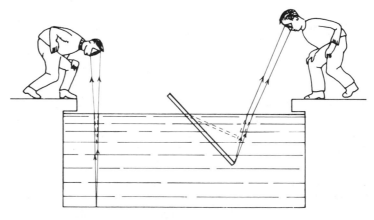

Fig. 6.15

Fig. 6.13. We will consider this colour effect again in Sections 6.11 and 6.25.

The shallow-looking swimming bath and the bent stick can easily be explained by our seeing a virtual image along the straight lines we 'know' light usually takes. Fig. 6.15 shows how these illusions arise, the size of the eye is exaggerated to make the mechanism clearer.

6.6 The reason for the bending

A similar refraction property was found in Section 5.13. The reason why water waves alter their direction is that they alter their velocity. Can this be the reason for light's refraction too? Velocity of light measurements are hard to perform for light travelling through air or empty space, let alone water or glass. Long optical paths and elaborate timing techniques are needed on account of the extremely high speed of light and the difficulties of achieving these in water, say, can be imagined. However, in 1850 Foucault did perform an experiment with water and the result showed a definitely reduced speed in water, which is just the right thing to account for the light waves bending the way they do on entering or leaving the water. Nowadays the speed of light can also be measured in glass or plastic and these too carry light waves at a slower speed than air.

6.7 Putting a number to 'bendability' – Refractive index

The amount of slowing which a transparent material causes compared with a vacuum or empty space can be judged from a quantity called **refractive index**.

$$\frac{\text{Refractive index}}{\text{of a material}} = \frac{\text{Velocity of light in empty space}}{\text{Velocity of light in the material}}.$$

Glass commonly has a refractive index of about 1·5, meaning that light travels at $\frac{2}{3}$ of its empty space speed in glass. Diamond reaches a value of 2 or more, i.e. light is slowed down by more than $\frac{1}{2}$.

From the way refractive index is measured it follows that a vacuum has a value 1 for it and other materials have values greater than one. The 'bendability' of a material is given by the extent by which its refractive index number exceeds 1. Air's refractive index is just greater than 1, but normally so close to it that the difference is ignored except for the most accurate work.

6.8 Refractive index and the angles

Taking advantage of more advanced physics than this book can cover there is a simple connection between the angles made by the light directions with the normals and the velocities – Fig. 6.16. It is:

$$\frac{v_1'}{v_2} = \frac{\sin i_1}{\sin i_2}$$

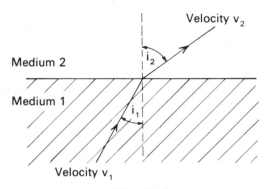

Fig. 6.16

This can be put in terms of the refractive indices n_1 and n_2 (Section 6.7):

$$\frac{v_1}{v_2} = \frac{v_0/v_2}{v_0/v_1} = \frac{n_2}{n_1}$$

where v_0 is the free space velocity. The combining of these two equations gives:

$$\boxed{n_1 \sin i_1 = n_2 \sin i_2}$$

This is by far the easiest way of remembering and using the relation between angles and refractive indices and is sometimes called **Snell's law** after William Snell, who discovered it as long ago as 1621.

Example 1
A beam of light falls on a glass block at an angle $i_1 = 30°$. The glass has a refractive index of 1·5, air of 1. What is the angle i_2 in Fig. 6.17?

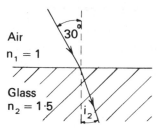

Fig. 6.17

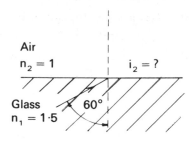

Fig. 6.18

Using the equation:

$$1 \times \sin 30° = 1·5 \times \sin i_2$$

$$\sin i_2 = \frac{1}{1·5} \times \sin 30°$$

$$= \frac{0·5}{1·5}$$

$$= 0·33$$

so $$i_2 = \underline{19°}$$

Example 2

The largest value that the sine of an angle can have is 1, so if we had the situation of Fig. 6.18 what would i_2 be? $n_1 = 1·5$, $n_2 = 1$, $i_1 = 60°$. Trying the equation this time gives:

$$1·5 \times \sin 60° = 1 \times \sin i_2$$

$$\sin i_2 = 1·5 \times \sin 60°$$

$$= 1·5 \times 0·866$$

$$= \underline{1·30}$$

There is no angle with a sine of 1·30. This means the light will have to bounce off the glass surface; there is no other possibility. When this happens we have **total internal reflection** of the light at the boundary.

6.9 Critical angle

In the last example 60° is easily big enough to make sin i_2 greater than 1, but what is the *smallest* angle that will do this? Clearly this happens when sin i_2 is just equal to 1. Fig. 6.19 shows the general situation for two materials of refractive indices n_1 and n_2.

$$n_1 \sin c = n_2 \sin i_2$$

For sin i_2 to be equal to 1 we must have:

$$\boxed{\sin c = \frac{n_2}{n_1}}$$

If $n_1 = 1.5$ (glass), and $n_2 = 1$ (air),

$$\sin c = \frac{1}{1.5}$$

$$= 0.67$$

so $\quad c = \underline{41\frac{1}{2}^\circ}$

The **critical angle** for air and glass is thus $41\frac{1}{2}^\circ$. Any light falling on the

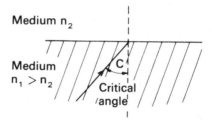

Fig. 6.19

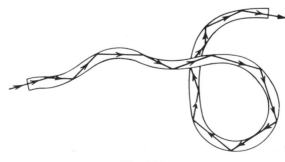

Fig. 6.20

boundary and travelling in glass towards the air at an angle bigger than $41\frac{1}{2}°$ will be totally reflected.

An interesting application of this effect, of growing importance, is so-called fibre optics. A beam of light can be trapped inside a solid rod of glass fibre (Fig. 6.20) by a succession of internal reflections all at angles greater than the critical angle. Using this technique photographs can be taken, for example, of the inside of a man's stomach by passing a flexible 'light pipe' down the oesophagus.

6.10 Refraction and squad marching

People can be made to bend just like light! A row of people link arms and practise marching in step at two speeds, one normal and the other at a much slower speed with shorter steps. A hall or outside area is needed with a convenient line to mark the 'boundary'. Fig. 6.21 shows a row *AB* approaching the line and as soon as any person in the row crosses the line he must march at the slow speed. The result is that the whole line slews round and finds itself marching in a different direction.

The opposite bending happens if the people speed up on crossing the line (Fig. 6.22) and even total reflection will happen if the angle is large and the change of speed appreciable – Fig. 6.23. Clearly some discipline in marching is needed to do this successfully, but it does illustrate the point very well – it is a change of speed which causes refraction.

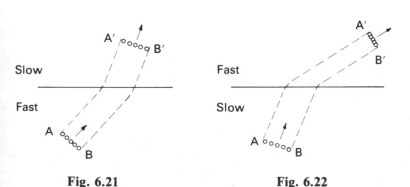

Fig. 6.21   **Fig. 6.22**

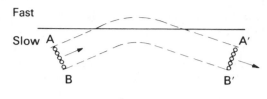

Fig. 6.23

6.11 Why different colours?

A narrow beam of white light can be split up into different colours by a prism – Fig. 6.24. Newton performed this experiment some 300 years ago. The inference is that white light is made of a mixture of many different colours and that a particular glass has its own refractive index for each colour. Another way of putting it is that refractive index is a function of the wavelength of the light, the shorter wavelengths having the higher refractive indices and therefore being bent more than the longer wavelengths. Colours and spectra are dealt with in more detail in Section 6.25.

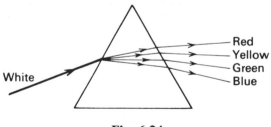

Fig. 6.24

6.12 What lenses do to a beam of light

A lens is a piece of glass or plastic whose section is like Fig. 6.25, the centre and the edges being of different thicknesses. There can be cylindrical or spherical lenses and their main property is that they alter a beam of light passing through them. A fan of light beams, made in a simple way shown in Fig. 6.26, can be directed in turn on each kind of lens.

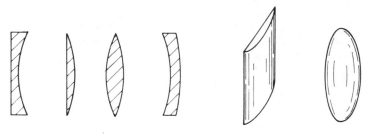

Fig. 6.25

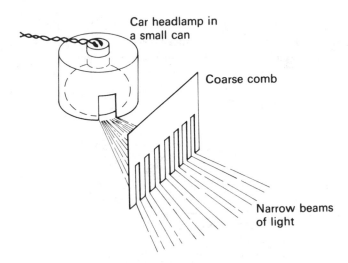

Car headlamp in a small can

Coarse comb

Narrow beams of light

Fig. 6.26

One kind of lens brings a fan of light beams to a focus (Fig. 6.27) and is called a **converging lens**. This kind is thicker in the centre than at the edges. The other kind of lens, thicker at the edges than in the centre, splays out the beam and is called a **diverging lens** – Fig. 6.28.

Usually we restrict our work with lenses to 'thin' ones, meaning lenses where the difference in thickness between the centre and the edges is much smaller (10 or 20 times) than the diameter of the lens. The reason for doing this is to simplify matters because if a fan of light beams is sent, for example, towards a thick converging lens, the outer

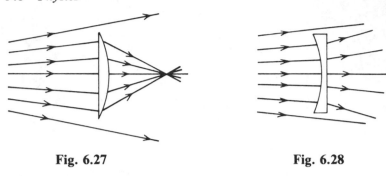

Fig. 6.27   **Fig. 6.28**

beams are not brought to the same focus as the central beams –
Fig. 6.29. A thick lens is of course more powerful than a thin one because
it focuses the beam in a shorter distance, but it suffers from this serious
defect of focusing which can spoil the image it may be used to form. In
practice when a strong lens is required it is often better to use a
succession of thin lenses, each doing a small part of the light-bending.

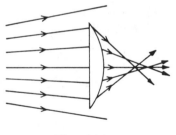

Fig. 6.29

6.13 Image formation

The main trouble with the pin-hole camera discussed in Section 6.2
is that the image it forms at the back of the camera is very dim,
necessitating a very long exposure time. An interesting thing happens
if more than one pin-hole is put in the front of the box – multiple
images are formed, one for each hole – Fig. 6.30. The solution to the
brightness problem would be to collect these separate images and
superimpose them. To do this the light would have to be converged,

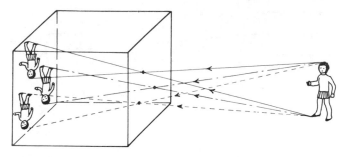

Fig. 6.30

just what a lens can do, but it will have to be a lens of just the right power to bend the light by exactly the right amount – Fig. 6.31.

There is now no need to limit the number of holes since, if it is the right lens, light going anywhere through it will be focused together. This will give perhaps a thousand-fold increase in brightness, a thousand-fold decrease in exposure time, and we have discovered the lens camera.

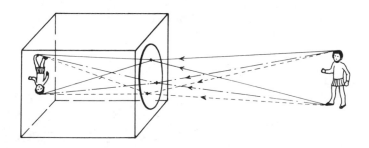

6.14 The power of a lens – its focal length

For quantitative work we need to be able to grade lenses according to their light-bending properties and a useful distance is the **focal length**. This is the distance from the lens to the focus point when using a *parallel* beam of light – Fig. 6.32. Clearly a long focal length means a weak lens, a short focal length a powerful one – Fig. 6.33. The point *F* is called the principal focus of the lens. A rough value for the focal

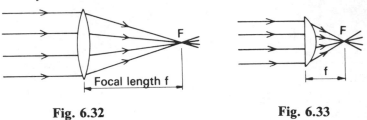

Fig. 6.32   **Fig. 6.33**

length of a converging lens is obtained by using light from a far distant point like a hill or tree or building. The light rays from each point on the distant object will meet the lens in almost parallel beams and the picture can be caught on a piece of white paper – Fig. 6.34.

Diverging lenses are more difficult in this respect. The focal length is taken as the distance from the lens to the point from which parallel light beams appear to diverge after passing through the lens – Fig. 6.35. Again a large focal length means a weak lens. It is not easy to find the focal length of a diverging lens directly; if a selection of converging lenses is available the simplest way would be to choose that one which just compensates for the diverging lens when they are held together. The focal lengths will then be equal.

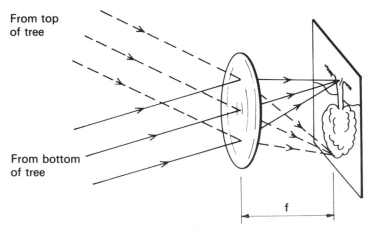

Fig. 6.34

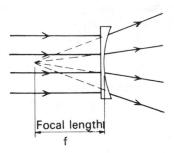

Fig. 6.35

6.15 Images and focal lengths

The general behaviour of converging lenses can be studied by using a lamp and a screen: the lamp should be bright but only allowed to shine through a small hole at the same height as the lens. Images of the hole

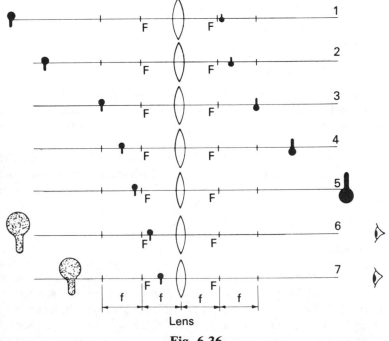

Fig. 6.36

can be formed on the screen. Suppose a rough value of the focal length has been found first, using a distant building.

With the hole a long way from the lens the screen catches a small, upside down picture of the hole very close to F – Fig. 6.36, line 1. As the hole is brought closer to the lens the screen has to move away a little and the image gets larger, though is still smaller than the hole itself, line 2. At line 3 the hole is two focal lengths away and so is the image, now grown to life size. As the hole comes closer, lines 4 and 5, the screen must be moved much further away to catch the now magnified picture.

At line 6 the hole is nearer than the principal focus, F, and now the screen cannot find a picture anywhere. But if an observer looks through the lens he sees a magnified picture, the right way up apparently, behind the lens, and again at line 7, but this time not so big or far away. These last two are examples of virtual images, met earlier with mirrors in Section 6.3. There is no possibility of putting these on a screen since no light goes to them, but they are images nonetheless and have important applications.

Optical instruments

6.16 The lens camera

The situation is that of line 1 of Fig. 6.36, an object a fairly long way away and its image close to F to keep the camera reasonably short. A small image is no handicap as enlargements can be made later if necessary.

Objects at different distances away can be dealt with by moving the lens nearer to or further from the film where the image must be focused. The picture on the film will be inverted. Lens cameras can focus all object distances from a few cm to the far horizon or even the stars by adjusting the lens–film separation.

6.17 The eye

This is another line 1 arrangement, with light focused by the cornea and lens on to the sensitive retina at the back of the eye. The relevant parts of the eye are sketched in Fig. 6.37. To cope with light arriving from different distances away the lens in the eye can be squashed or relaxed by muscles round its rim to make it more or less powerful. The

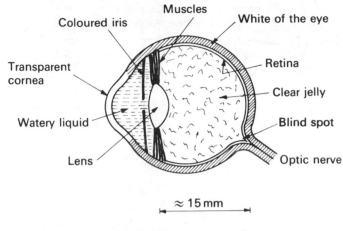

Fig. 6.37

normal range of distances which average adults can accommodate is from the far distance ('infinity') up to about 25 cm in front of the eye. Most of the bending is in fact done by the cornea, the lens inside just adds the fine adjustment necessary for clear vision.

6.18 Spectacle lenses

The figures mentioned in the last section give the average range of distinct vision for the population as a whole, but many people suffer from long- or short-sightedness which means they need lenses to help the eye to focus properly.

Longsighted people have difficulty in making the eye lens sufficiently powerful to focus on nearby objects – Fig. 6.38. It may be that the nearest distance of distinct vision is a metre. A lens is needed which will do some of the bending for the eye and which will make, at a metre distance, a virtual image of an object brought up to 25 cm. This corresponds to line 7 of Fig. 6.36. Longsighted people need converging spectacle lenses.

Shortsighted people suffer from the opposite trouble. Nearby objects present no problems, in fact their near point may be down to 10 cm instead of the normal 25 cm. It is the distant objects which cannot be accommodated this time – Fig. 6.39. The eye lens is always too

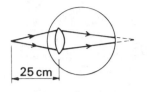

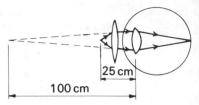

Longsighted eye cannot focus on objects close to the eye

Correcting lens for a long-sighted eye of near-point 100 cm

Fig. 6.38

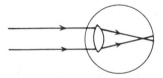

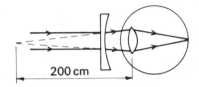

Shortsighted eye cannot focus light arriving from a distant object

Correcting lens for a short-sighted eye of far-point 200 cm

Fig. 6.39

powerful for the distant object and a lens is needed which will diverge the light slightly before it enters the eye. Shortsighted people need diverging spectacle lenses.

6.19 Magnifying glasses

A watchmaker's magnifying glass enables him to study the tiny mechanism inside a watch. It is held close to the eye and forms a much magnified upright image for him as in line 6 of Fig. 6.36. (Obviously the image must be in front of the eye to be seen and the only other case of high magnification would be line 5, but this arrangement would give an upside down image behind the man's head!)

What is happening to the light directions is shown in Fig. 6.40. In effect such a lens brings the eye's near point down to a few cm. The object to be viewed should be placed just inside the lens's focal point

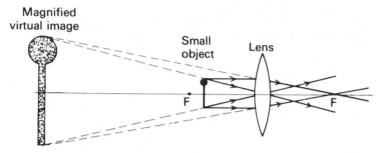

Fig. 6.40

and the lens chosen to give a highly magnified picture at least 25 cm in front of the observer's eye.

6.20 Microscopes

When faced with the need to obtain very high magnifications it is far simpler to do it in stages rather than in one big boost, and the common microscope uses two stages to achieve its high magnification.

Referring again to Fig. 6.36, a line 5 arrangement is followed by a line 6 and the overall magnification can be in the hundreds or more. The small object lies just outside the focus of a short focal length lens; this gives an image several focal lengths away, upside down and magnified. This image becomes the object just inside *F* for a second lens used as a magnifying glass. The final virtual image must be at least 25 cm from the eye of the microscopist.

The eye should be placed so that it receives light from all over the lens aperture and all over the image. Fig. 6.41 shows this position and also the light paths through the instrument.

Real microscopes of course have several lenses doing the jobs of the single ones shown in the drawing, but the result is merely a better quality image, still upside down compared with the original object. For highest magnifying power both lenses need to be of short focal length, the objective lens very short indeed, and separated by many times its focal length from the eyepiece.

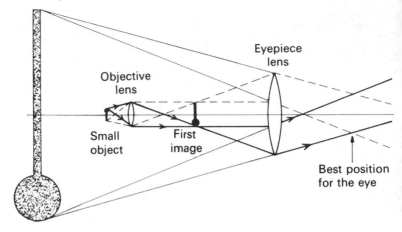

Fig. 6.41

6.21 Lens telescopes

A different situation confronts us with making a telescope. The object is a long way off so we are forced to start off with a line 1 position – Fig. 6.36. There is no possibility of getting the first lens close to the object. If there were, why make a telescope?!

The line 1 position presents us with a small image very close to F, upside down, but really there. Contrast this with the microscope – the first lens there does a lot of magnification itself, but the telescope has to begin with a much diminished image. We will want this first image to be as large as possible, though, so that the final image is large, and this means that the first lens must have a long focal length. The snag is that a very long focal length will mean an unwieldy instrument (Fig. 6.42) and for hand use a little over 60 cm is the limit.

The small image produced by the objective lens must now be magnified considerably just like the first image in the microscope, another line 6 position. The two lenses give a magnified virtual picture as shown on Fig. 6.43, but it is inverted again. Such an arrangement is called an **astronomical telescope** since, obviously, for looking at a dot of light in the sky it does not matter much if it is upside down.

If a telescope is required to give an erect picture something different is needed. There are three ways of doing this. The obvious one is to turn the intermediate image over before the eyepiece lens magnifies it.

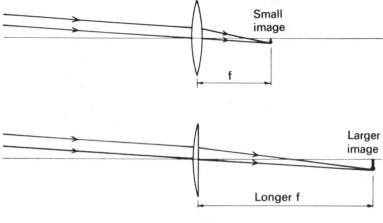

Fig. 6.42

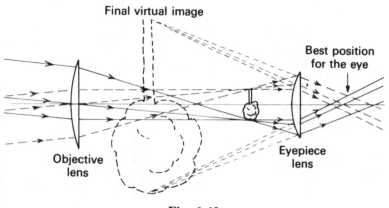

Fig. 6.43

Another lens can do this (Fig. 6.44), but it does add to the overall length of the telescope. The erecting lens is used in a line 3 position of Fig. 6.36 because this adds least to the length and does not do any magnification. Three lenses used like this comprise a **terrestrial telescope**, used for viewing distant objects which must be seen the right way up. Coastguards' telescopes and some gunsights are made like this.

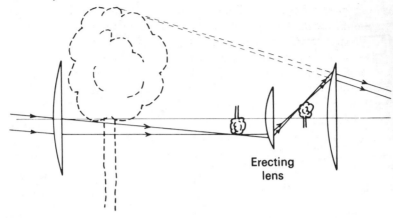

Fig. 6.44

A more commonly used way of erecting the intermediate image is to do it with a pair of prisms. The prisms are really used merely as mirrors, so two are needed to give a complete inversion of the image. The main advantage is that the length of the instrument can be greatly reduced as the light path doubles back on itself. Fig. 6.45 shows the

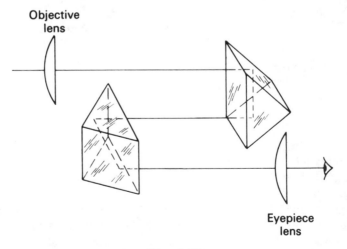

Fig. 6.45

arrangement with just one light path traced through the centre of the system. Binoculars use prisms in this way, being telescopes, but with the physical length reduced to a little more than a third.

The earliest form of telescope used a quite different way of making an upright image. Instead of using a converging lens for the eyepiece a diverging one was used – Fig. 6.46. It has to be placed in front of the image made by the first lens, but it does make an upright image of it. The total length of this **galilean telescope** is rather less than the astronomical one, but one disadvantage is that its angle of view is rather restricted. It finds a use in opera glasses, some gun sights and cheap binoculars.

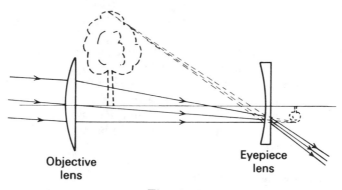

Objective
lens

Eyepiece
lens

Fig. 6.46

6.22 What do optical instruments really magnify?

It is a well known effect that distant objects appear small; the further away they are the smaller they appear. The sun, for instance, seems to be almost exactly the same size as the moon. When it comes to making an image appear large there is not much point in making it big if it is also positioned at a great distance from the observer. What we want is the image as near as comfortably possible and as large in that position as the lenses will allow. The size of an object or image is decided mainly by the size of the stimulated area on the retina of the eye and Fig. 6.47 shows that it is the *angle* filled by the object which decides this.

The job of a magnifying instrument like a microscope or telescope

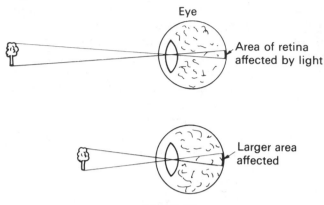

Fig. 6.47

is to make an image which fills a larger angle than the original object. After all, there is no real chance of making an image of the moon larger than the actual moon, but we can make a small image so much closer than the moon itself that it fills a larger angle and appears to be magnified. The microscope has a clearer job, simply to magnify diameter since both object and image can be brought up to the eye's closest comfortable viewing distance.

6.23 Resolving power – detail viewing

Telescopes and microscopes have a bigger job to do than merely making a magnified image. If that image is of poor quality there is no point in magnifying it. What we want to see is magnified detail, the tiny parts of some biological specimen or the exact nature of Venus's cloud covering. This is not the same thing as magnification and it depends on different factors. The full treatment of this subject is beyond the scope of this book, but the main results are worth noting.

Telescopes
The important thing here is the size of the objective lens – its diameter or aperture. In fact telescopes are described like this: a 12 cm refracting telescope, or a 50 cm one. The larger the light-collecting area the greater the detail which can be seen. For this reason observatories build large telescopes with apertures many metres across to see very small stars deep in space.

Microscopes

To see great detail in a microscope image the object lens must accept a wide angle of light from the object. This means the object and lens have to be close together, but, remembering that the object lens does a line 5 job from Fig. 6.36, they must still be more than the focal length apart. Very short focal length lenses are therefore needed, may be as low as 2 mm. Another way of improving the resolving power of a microscope is to immerse the object in an oil of high refractive index.

Yet another factor involved with resolving power is the wavelength of the light used. Light behaves like a wave motion with very short wavelength; in fact it is its wave behaviour which causes there to be a limit to the observable detail. The point is that for both microscopes and telescopes a reduced wavelength gives a greater resolving power, so ultra-violet and X-ray microscopy has developed to take advantage of the fact. Electron-microscopes take the same point a stage further because although electrons generally behave like particles they do have wave properties and can be focused and bent very much like light. Telescopes to 'see' in the ultraviolet or X-ray fields have to be carried above the atmosphere by rockets since these wavelengths do not penetrate to ground level in sufficient strength.

6.24 Reflecting telescopes

When telescopes of high magnification and high resolution are being made, the first requirement means a long focus lens and the second a large aperture. In construction this means a very big lens, thick in the centre containing glass of considerable weight, yet able to be supported only around the rim. Glass disperses and absorbs some of the light going through it and is very hard to manufacture with uniform refractive index. Being really a liquid, glass 'flows' over a period of time. Clearly it would be better to avoid large lenses and the answer is to use concave mirrors instead. These suffer from none of the defects mentioned for lenses and can even be made parabola-shaped to give a really good image.

There are two optical systems often used for reflecting telescopes drawn simply in Figs. 6.48 and 6.49. All the largest telescopes use mirrors to collect the light and may be as large as 5 m across.

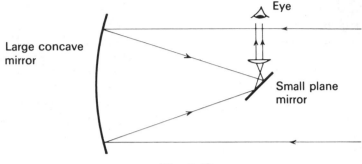

Fig. 6.48

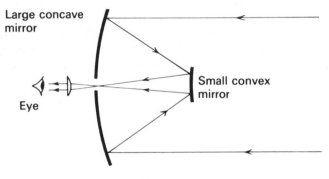

Fig. 6.49

6.25 Colours and spectra

In Section 6.11 it was decided that while light was made up of a whole range of colours, the rainbow colours, each bent slightly differently by passing through a prism. Blue and violet light is bent more than the other colours. The simple arrangement drawn, then, does not give a very good spectrum, a better arrangement is shown in Fig. 6.50 using a lens. If the screen is translucent it can be viewed from behind and a patch of light is seen shading imperceptibly from one colour to the next from red to violet.

Another way of making a spectrum is to pass the light through a diffraction grating instead of the prism. This is a piece of glass or clear plastic on which are ruled thousands of narrow 'lines' making an

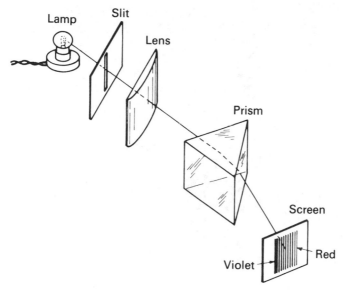

Fig. 6.50

extremely fine comb of many slits. The light waves spread out through each slit and from Fig. 6.51 it is seen that in certain directions the wave crests are all in step. In these directions a bright light will be seen, and each colour having a slightly different wavelength will give brightness in slightly different directions, producing a spectrum of colours. The advantage of the grating over the prism is that it can be made very thin and absorb hardly any light, and also that it spreads the colours out much more evenly than a prism. The order of the colours is reversed compared with a prism, too, and there are several spectra – Fig. 6.52.

The important point about spectra is that they are characteristic of the thing making the light and can be used to identify substances, and even to analyse quantitatively quite complex light sources such as stars. With an ordinary filament lamp all colours are present and we get the familiar rainbow effect, but with light from sodium street lamps or neon signs or the sun there are very definite colours present or missing which serve to identify the sodium or neon or the elements present in the sun's atmosphere.

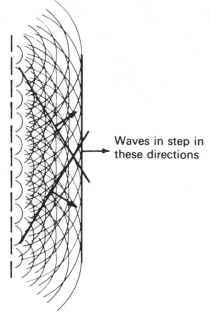

Fig. 6.51

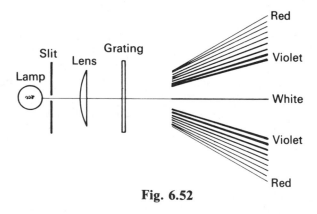

Fig. 6.52

6.26 Filters and colour mixing

If a red filter is placed in the beam of light used to make a full spectrum the result is that only the red colour gets through. Instead of a wide

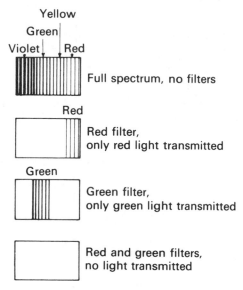

Fig. 6.53

spectrum spanning all the colours, only a narrow red portion is seen – Fig. 6.53. Similarly, a green filter only lets the green part of a spectrum through.

Suppose a green filter and a red filter are placed in the beam? Each one transmits only one colour which the other one does not transmit, so there can be no light transmitted in the end.

Suppose we have two separate lights, one with a red filter and the other with a green filter, arranged so that their beams overlap. What colour is seen then? Fig. 6.54 shows the arrangement and the result is yellow! Simple rules for adding colours can easily be obtained from experiments like this:

red	+ green	= yellow
red	+ blue	= magenta
green	+ blue	= turquoise
red	+ green + blue	= white
red	+ turquoise	= white
green	+ magenta	= white
blue	+ yellow	= white

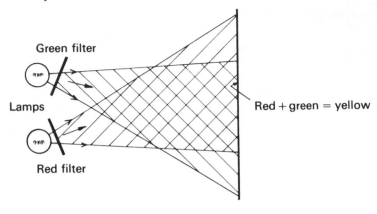

Fig. 6.54

The importance of this scheme lies in its application to colour photography, stage lighting and colour television. Each of these uses a three-colour system to produce the effects of full colour by mixing them in the correct proportions. Colour television, for instance, uses a cathode ray tube with three separate electron beams and a composite phosphor having three different kinds of colour emission. Grids close to the screen inside the tube make sure that each beam excites only those phosphors of one colour and by altering the relative intensities of the electron beams the effect of full colour pictures can be reproduced.

Note:
1 The rules for mixing coloured lights are not the same as those for mixing paints or pigments. This latter process is one of subtraction, each pigment *removes* some colours from the light falling on it until a complete mixture of pigments makes black. With the coloured lights we are *adding* different colours together.

2 The position of yellow in this scheme is awkward because the brain registers yellow for two quite different events. Red and green lights falling simultaneously on the retina produce the visual sensation of yellow. But there is a yellow too, a colour in its own right, emitted by sodium vapour street lamps, which produces the same visual sen-

sation as the red/green mixture. It is the red/green yellow which features in the above colour-mixing scheme.

6.27 What is light?

We now know a great deal about how light behaves. It travels in what we call straight lines, it bounces off shiny surfaces with equal angles, it changes its direction when crossing from one transparent material to another, it can be focused by lenses and used to form magnified images, it will show interference effects, it will bend round very small obstacles, it travels unbelievably fast.

With all this information one might expect us to know not only how light behaves but also what it is. It was remarked at the end of the first chapter, however, that the scientist's job is not to decide why natural phenomena *are*, but rather to describe their interactions in the light of some model or other. There have always been two models for light and the battle between them raged (literally at one time) from 1670 to around 1850. The models were a particle stream and a wave motion. The reader may find the historical development of the two ideas interesting; it certainly illustrates the point that scientists are just as selfish and intolerant as the rest of the community. An account will be found in A. E. E. McKenzie's books, *The Major Achievements of Science*, Vols. I and II.

Around 1850 the matter was thought to have been settled in favour of waves by the arrival of Clerk Maxwell's theory describing light in terms of electro-magnetic waves, even predicting correctly the speed at which they ought to travel. The main experiment which swung ideas decisively towards waves was the two-slit interference achievement of Thomas Young in 1802, Section 5.12.

Towards the end of the nineteenth century, however, and at the beginning of the twentieth some new experimental observations arose which were not satisfactorily explained by the wave theory of light. These experiments involved the brightness and colour of light radiated by hot objects and by electrically excited gases at low pressure, and the emission of electrons by many metals when light is shone on to them (the photoelectric effect). The result of these observations was that the old idea of particles was revived in a new form; if light energy travels in clumps or separate bits – called **photons** – the awkward experimental results could be explained quite beautifully.

The modern view is that light behaves like both waves and particles depending on our choice of conditions. Both ideas are taken to be valid, complementary models of light, illustrating exactly the earlier point that scientists cannot say with certainty how the natural world is made, only how it behaves. When dealing with mirrors, lenses and small holes light behaves as if it were waves of some sort, but when dealing with atomic processes like absorption and emission of energy it behaves as if it were made of separate chunks which we call photons. And there the matter rests.

6.28 As fast as we can go?

In 1905 Albert Einstein published what is called the Special Theory of Relativity. This is a mathematical work, but it is based on the strange observation that it does not matter how the speed of light is measured one always gets the same value. There is no effect like wind which alters the measured speed of sound in different directions when the wind is blowing. The absence of the wind effect is odd because all other wave phenomena show it. Einstein makes this fact the cornerstone of his theory and from it there follow several peculiar results:

Moving objects appear to shrink in the direction of their movement relative to stationary observers.

Moving clocks appear to lose time compared to a stationary clock.

Moving masses appear to increase their mass compared to their rest values.

Energy and mass are interchangeable, merely different aspects of the same physical phenomenon.

The velocity of light in vacuum is the maximum speed at which any form of energy (or mass) can travel.

These outlandish predictions become even more weird when you realise that 'moving' and 'stationary' can equally well apply to either observer. In Fig. 6.55, A sees the shrinking table and slowing clock of B, but B says just the same thing of A's table and clock!

All these effects have in fact been observed in experiments. The energy/mass one in particular has manifested itself in our nuclear bombs and power stations. Since the theory agrees with what we can measure it has been accepted as a *good* theory, a fair description of the real world. The point about the maximum permissible speed being

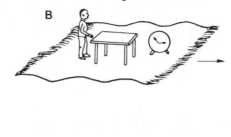

Fig. 6.55

that of light in free space (nearly $300\,000\,000\,\mathrm{m\,s^{-1}}$) is also accepted – until new evidence presents itself to the contrary and that has not happened yet.

(The relativity effects mentioned above are not generally noticed in everyday life because the speed of light is so large compared with our usual speed of movement. To get even a 1 % change in size or time or mass requires a relative speed around $\frac{1}{7}$ that of light!)

7

Charging About – and Current Affairs

7.1 + and − What's in a name?

In the land of Rojotania red-haired people were extremely rare. If a baby was born with red hair it was a matter for great rejoicing and the child grew up to be a specially favoured member of the Rojotanian society. The king of the land was very lucky because his first son had red hair and in his joy the king decreed that in future throughout the land all red-haired children should be called Roj. The people of the land were happy with their king and gladly complied with his decree and the habit remained for several hundred years.

Unfortunately the Rojotanians were not very meticulous about keeping records and after many centuries some of the people began to wonder why all red-haired children were called Roj. Why not call them John or Jane or Peter or Ann? No real reason could be found and some of the inhabitants were rather worried about the custom. However, wise counsel prevailed and they realised that it was simply a matter of convenience – a decision had been taken which was quite arbitrary, but which had become over the period of time an ingrained habit which everyone observed, and so to this day the name Roj infallibly labels a red haired Rojotanian.

The point of this legend is to illustrate the situation concerning the names positive and negative in electricity. Why these names? No real reason, but we have decided upon them and will have to stick to them now. A long time ago men found that certain materials rubbed with cloth or fur would pick up small objects. They found there were two kinds of 'charge' and so they labelled them with convenient names: + and −. To say that an electron is negatively charged means it has the same electrical property as a piece of ebonite has when rubbed with fur! Protons are said to be positive because they behave like glass rubbed with silk. There are no magical reasons for the names – just that a long time ago someone decided – and that was that.

7.2 The electron – a short history lesson

About 600 BC a Greek philosopher, Thales, found that amber which he had rubbed with fur could pick up small bits of straw. Not much happened from then until about 1700 when Newton imagined electricity to be a 'weightless fluid'. Benjamin Franklin mentioned the idea of 'electrical particles' about 1750, but it was not until the nineteenth century that real progress was made. Faraday did some important work on the passage of electricity through liquids and concluded that atoms were 'in some way endowed with electrical powers'. Weber in 1871 and Helmholtz in 1881 took up the particle idea and the latter considered positive and negative electricity to be particulate; but up to almost the end of the century there was no direct evidence.

The development which eventually clinched the matter was the study of cathode rays, begun seriously around 1855 and culminating in the work of J. J. Thomson in 1897. He showed that whatever comes from the cathode (negative electrode) when an electric current is passed through a low pressure gas behaves just like a stream of particles carrying negative charge. Thomson also showed that compared with the lightest known atom these particles either had very much more charge or very much less mass (Section 7.6). The latter turned out to be the case and was decisively shown to be so by an elegant set of experiments of R. A. Millikan early in the twentieth century, the details of which we will discuss later (Section 9.5). The point of Millikan's experiments is that he found electrical charges exist only in multiples of a certain minimum charge which he could calculate. This is the charge carried by the electron, the particle in Thomson's cathode ray experiment.

7.3 Two units to trust

Electric current is measured in **amperes**, or amps for short. Everyone is familiar with these through 13 A plugs and fuses, ammeters on cars, etc., yet very few people know just what an amp is. This matters little of course, and we will be content to follow the habit and accept ammeters marked for us in amps to measure electric currents.

A current, though, is a flow of something. We know about water currents, air currents and traffic currents, and it is clear just what it is that is moving – water, air and cars. What flows in an electric current? To say 'electricity' does not solve anything. The answer is electric **charge**, the thing the ancient Greeks knew about by rubbing. What do we measure charge in? – **coulombs**. How are amps and coulombs related to each other? In an electric current we have a steady flow of charges round the circuit and could count how many coulombs go past any point in a second. That tells us the number of amps.

> **1 amp = 1 coulomb per second**

A current of 5 amps means a flow of 5 coulombs every second. Sometimes we use small currents and use milliamps or even micro-amps, symbols mA and μA, being respectively $\frac{1}{1000}$ and $\frac{1}{1\,000\,000}$ of an amp. (In terms of electrons moving round the circuit, 1 amp means about 6 250 000 000 000 000 000 ($6 \cdot 25 \times 10^{18}$) electrons per second!)

7.4 The volt – electromotive force

Another commonly used electrical quantity is the volt. We have 240 V mains pumped into our houses; cars use a 12 V battery; transistor radios need a 9 V battery; torches often use $1\frac{1}{2}$ V or 3 V. Overhead power cables carry electrical energy at 132 000 V or 400 000 V and main-line electric locomotives operate from 25 000 V. Here again, not many folk bother about how a volt is made nor do they need to, but this time we *do* need to know exactly what 1 V means.

Consider a simple electrical circuit made up of a battery, two wires and a lamp – Fig. 7.1. Something is flowing round the circuit. If an ammeter is placed in the circuit it always gives the same reading no matter where it is inserted – on the left of the lamp or on the right or even between the two cells of the battery. Everywhere there is the same flow, the same number of electrons per second passing any point. What keeps the electrons moving? The battery of course. Where does the light and heat energy being sent out of the lamp come from? Again, the battery. Where does the battery get the energy from? From chemical changes which take place in the materials making up its

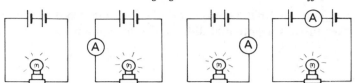

Fig. 7.1

plates. The battery then is a supplier of energy (until its store of chemical energy runs out) and we measure its ability to move electrons – or its **electromotive force** – in **volts**. If the battery transfers 1 joule of energy from chemical into electrical for every coulomb it sends round the circuit, we say its e.m.f. is 1 volt. A 12 V car battery supplies 12 joules of energy for every coulomb sent round the car's circuit. Putting this into symbols, if W joules are involved when q coulombs go round the circuit, the battery has an e.m.f. of V volts where:

$$V = \frac{W}{q}$$

or, **1 volt = 1 joule per coulomb**

7.5 The volt – potential difference

Quite often we wish to consider only part of a circuit or to take charges from one conductor to another without going round a complete loop. We use the same idea to find the difference of electrical state – called potential difference – between the two points. If W joules

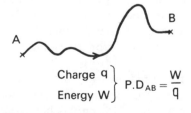

$$P.D_{AB} = \frac{W}{q}$$

Charge q
Energy W

Fig. 7.2

of energy are needed to take a charge of q coulombs from one point to another there is a potential difference of V volts between the two points, $V = \dfrac{W}{q}$ (Fig. 7.2). It does not matter which path we take – why not?

It is easy to see how this fits with a complete circuit discussed in Section 7.4 above, since that is made of a series of trips from one point to another. The energy needed for a complete loop is the sum of the energies for each part of it.

7.6 The earth – zero potential

From what we said just now about volts all we have is a measure of the difference between the potentials at two points, no hint of the actual potential at any point. What we need is a place where the potential is zero. There are two possibilities – a theoretical place so far away that all electrical influence is negligible, that mythical region called 'infinity'; alternatively, a practical place that will do, to all intents and purposes, as a stable electrical container. The position is similar to the need to choose a zero for height measurements to enable maps to be drawn accurately – a quite unobtainable zero called 'mean sea level' is used. In electricity we choose the earth to be our zero for potential measurements, on the assumption that it, like the sea, is a large enough reservoir that small additions or removals in practice make no discernible difference to its electrical state.

7.7 Electrical containers – capacitors

A litre of water put into a tall, thin container will fill it to a great height. The same volume put into a large tank will be lucky to make a puddle a few millimetres deep – Fig. 7.3. Similarly if a charge of, say, $\frac{1}{1000}$ of a coulomb is placed on a certain conductor its potential could rise by 1000 volts compared to the earth; if placed on a different conductor it may cause only a 10 volt rise. Different conductors have different 'storage abilities' – the property is called **capacitance** and the objects **capacitors**.

Capacitance is measured by the ratio of charge to potential difference. If 1 coulomb causes a 1 volt rise we have a 1 farad capacitor. This turns out to be impossibly large in practice and common sizes are microfarads (micro means millionth) or even micro-

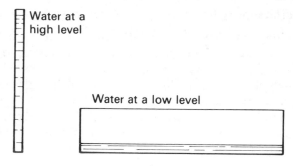

Water at a high level

Water at a low level

Fig. 7.3

microfarads, also called picofarads. In symbols:

$$C\,\text{(farads)} = \frac{q\,\text{(coulombs)}}{V\,\text{(volts)}}$$

7.8 Charging and discharging

The connection between charge and current can be neatly shown using a capacitor. A large capacitor, say 100 μF, is connected with two sensitive ammeters to a high voltage generator. On switching on, the ammeters flick showing that a current flows for a short time – Fig. 7.4. When the connections are removed from the voltage source and joined together the meters flick in the opposite direction. While charge is flowing a current is registered, perhaps quite a large one for a short while, and both meters move in the same direction.

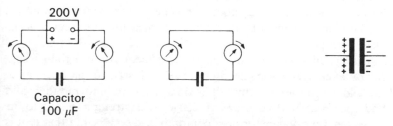

200 V

Capacitor
100 μF

Fig. 7.4

Fig. 7.5

7.9 A warning – lazy language

The capacitor in the charged condition is often depicted as in Fig. 7.5, with + charges on one plate and − charges on the other, and we even talk about + and − charges 'moving'. The reader ought to be aware that it is only electrons in conductors which are free to move and that a + charge is really the absence of electrons. What the voltage supply can be said to do in Fig. 7.4 is to remove some electrons from one side of the capacitor and put them on to the other – that is why the meters both move in the same direction.

7.10 What are capacitors used for?

The theory of these devices is beyond the scope of this book, but capacitors find so many uses that we should know some of these without a knowledge of the reasons why they work being important.

(*a*) A capacitor is used to reduce the sparking across the points of a car's distributor.
(*b*) Capacitors are used extensively in radio and television circuits – they will conduct alternating current but not direct current.
(*c*) A variable-capacitance capacitor is used to tune radio receivers so that only one station is heard at a time.
(*d*) Small induction motors in fan heaters or hair dryers need a capacitor as a phase shifter.

Fig. 7.6 shows a selection of differently made capacitors.

7.11 Electric fields

On earth a lump of 1 kg is pulled downwards with a force of 9·8 newtons – Fig. 7.7. The earth has a gravitational 'strength' of 9·8 N kg^{-1}, the moon only about 1·6 N kg^{-1}, empty space nothing at all. The force felt by a mass of unit size measures the gravitational 'field' strength.

The idea of a *field*, a region where the influence of something is felt, can be taken over wholesale into electricity. Instead of putting a kg in the field to test the strength of it, we imagine a coulomb put there and find the force it experiences. Electric field strength ought to be measured, then, in newtons per coulomb, and so it is, but that unit is identical with volts per metre (because a volt is a joule per coulomb

Fig. 7.6 Capacitors

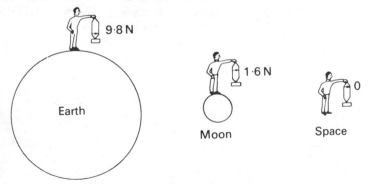

Fig. 7.7

and a joule is a newton-metre). The latter unit is used to show that field strength is like the steepness of a hill – the bigger the gradient the bigger the effect. Unlike gravity, though, we can get pushing and pulling forces between charges, so we have to use a + charge of 1 coulomb as our test object.

A field of 1 volt per metre (or 1 newton per coulomb) exerts a force of 1 newton on a charge of 1 coulomb placed in it. A field of E volts per metre exerts a force of E newtons on 1 coulomb, and therefore qE newtons on q coulombs. So we can write the force, F, as:

$$F \text{ (newtons)} = q \text{ (coulombs)} \times E \text{ (volts per metre)}$$

or

$$\boxed{\textbf{Force} = \textbf{Charge} \times \textbf{Field}}$$

7.12 Electric field patterns

A useful way of depicting a field of any kind is to draw a set of lines to represent it. The direction of the lines shows the field direction (i.e. the force direction felt by the test object), and the closeness of the lines indicates roughly the strength of it. A strong field will have densely packed lines, a weak field relatively few widely spaced ones. Fig. 7.8 gives some examples. The arrows show which way a + charge would move.

If the lines originate from or terminate on a conductor they will make a 90° angle with the surface. This must be so because in a conductor the electrons are free to move about within the metal and a field line shows in which direction a force acts on a charge at that point. If there were any component of this force along the surface the

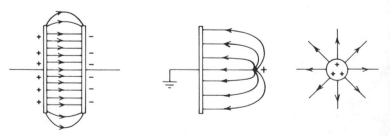

Fig. 7.8

charges would move until the force, and hence the field, were perpendicular to the surface. This means that conductors of irregular shape will tend to cluster their field lines round the most curved bits – Fig. 7.9, and therefore that charges will congregate at the points to give the densely packed lines and the strongest regions of field.

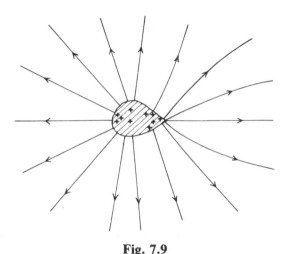

Fig. 7.9

7.13 Lightning conductors

Thunderstorms produce clouds which become electrically charged. The mechanism by which this happens is complex, but lightning occurs when the field built up by these charges becomes sufficient to break down the air into ions (charged molecular fragments). This occurs in dry air under normal conditions when the field exceeds 3 million volts per metre, but can be considerably less in humid conditions. The great energy carried by a charged thundercloud all released in one violent discharge can cause catastrophic damage.

If the energy could be released gradually and silently the damage could be avoided. Tall buildings often carry lightning conductors, thick copper cables running up from a good foundation in damp earth and ending in sharp spikes at the top. The field between the spike and charged cloud is drawn in Fig. 7.10 and near to the spike can be large enough to ionise the air. These charged fragments will be attracted

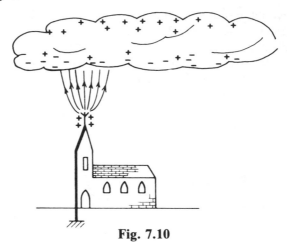

Fig. 7.10

to the cloud and slowly discharge it. The action depends on the stronger-than-normal field around the pointed end of the lightning conductor.

7.14 The Van de Graaff generator

A device which uses this peculiar action of pointed conducting objects to concentrate their charge on the end is the electrical generator, invented in America in 1931 by Van de Graaff. In the simplest design (Fig. 7.11) an endless insulating belt carries charge up to a large hollow conducting sphere. The charge is 'transferred' from the lower points to the belt by the high-field ionisation mechanism described in the last section. It is collected at the top by another set of points and gradually the sphere becomes more and more charged until its voltage reaches maybe half a million volts. The maximum voltage attained depends on the size of the sphere, the cleanness and smoothness of its surface and the breakdown field of the air around it.

These machines are used as a first stage for accelerating charged atomic particles like electrons and protons.

7.15 Flow of charge = current

The Van de Graaff generator can be used to show clearly that electric currents really are only charges in motion. If an earthed sphere is

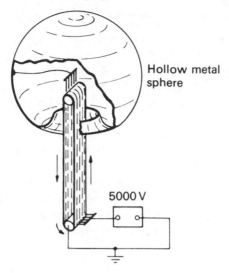

Fig. 7.11

placed near to the charged top sphere of the generator a spectacular spark (lightning flash) joins the spheres when the field between them is enough to ionise the air – Fig. 7.12.

If a very sensitive ammeter is placed in the lead from the small sphere it 'kicks' every time there is a spark. If the sparks occur frequently (put the spheres closer together) a steady deflection is obtained on the ammeter. The current is only microamps, but it is an ordinary electric current.

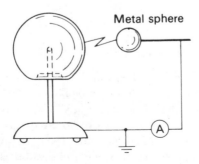

Fig. 7.12

7.16 The first rule about currents

Students are sometimes surprised when an arrangement like Fig. 7.13 is set up. All the ammeters show the same reading, yet it is a common belief that 'electricity' spurts out of the + terminal of the battery, gets 'used up' in the lamp and just manages to stagger back into the – terminal! Remembering that current is a flow of charge there can be no doubt that as much goes into any part of the circuit as comes out, otherwise there would be a build-up of charge somewhere or a serious lack of it. In a circuit like this, a **series** circuit, the current everywhere must be the same.

It follows from this that when we have a branching or **parallel** circuit the currents will divide along the branches, but at any junction the total arriving at it must equal the total leaving. The ammeter readings in Fig. 7.14 are taken from an actual circuit.

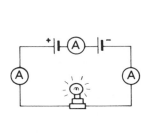

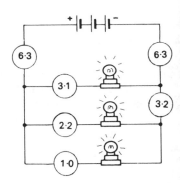

Fig. 7.13   **Fig. 7.14**

7.17 Which way do currents flow?

The Rojotanian legend of Section 7.1 shows how arbitrary certain names are and here is another decision taken a long time ago which we still accept. We **say** that electric currents flow from + to − outside a battery and usually put arrows on circuit diagrams showing that direction. This notwithstanding more recent knowledge that what actually moves in most circuits is electrons and these, being negative, will have to go the other way round!

7.18 The idea of resistance

If we connect a 2 V lead accumulator to a torch bulb the bulb filament gets hot, white hot in fact, and an ammeter may read 0·3 amp. If the battery is joined to a small chunk of carbon (a soft pencil lead will do) the reading may be 1 or 2 amps – Fig. 7.15. Connected only through thick copper wires we will get maybe 30 amps – and incidentally ruin the accumulator! Why is it that the same accumulator delivers such different currents? Why does it need joules of energy to send electrons through wires anyway?

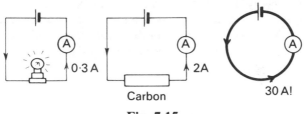

Fig. 7.15

The answer is that the electrons that do move have to thread their way through a forest of atoms or ions essentially fixed in place in the solid material they constitute. Only one or two electrons are available for conduction of electricity from each atom anyway. It is like a pin-table with balls rolling down. They accelerate for a short while, then suffer a collision, accelerate again and so on. The closeness of the pins will decide how quickly the balls get through the maze. Fig. 7.16

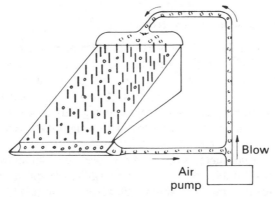

Fig. 7.16

represents a model of the circuit – the air pump does the job of the accumulator and the pins offer **resistance** to the flow of the balls.

7.19 Measuring voltage – joules per coulomb

The circuits shown so far containing ammeters make it clear that the meter has to be made part of the circuit – the current flows through it and causes the pointer to move. Voltmeters are used in a different way. We measure currents *through*, but potential differences *across* components. The voltmeter is added on to the circuit and measures how much energy is needed to send a coulomb between the two points it is connected to – as in Fig. 7.17. We might find 4 V across the dim bulb and 8 V across the brighter bulb – it is harder to take a coulomb through the brighter bulb, that is why it is brighter.

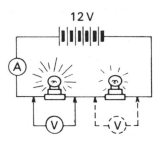

Fig. 7.17

7.20 Ideal ammeters and voltmeters

Looking again at Fig. 7.17 tells us what voltmeters ideally should not do. We do not want them to alter the state of the circuit when they are connected to it, we want *no* current to go through them! Most voltmeters do take a small current (mA perhaps), but if it is much smaller than the main circuit current that is good enough. If necessary voltmeters which operate in a different way and take no current at all can be found and used.

Suppose the voltmeter is connected across the ammeter in Fig. 7.17. If it reads at all it tells us how much energy is needed to operate the ammeter. Ideally, we want this to be zero; the ammeter shows the current flowing and should not need joules to make it work. Again, most ammeters do require a very small amount of energy to operate their mechanism and if this is negligible compared with other parts of the circuit we are satisfied.

Another way of putting the requirements about meters is that **ammeters** require a low resistance, ideally zero; **voltmeters** ought to have a high resistance, preferably infinitely high.

7.21 Ohm's Law – the Ohm

Thinking back to the pin-table idea of Fig. 7.16, there will be two things which determine the rate of fall of the balls. The closeness of the pins was mentioned before, but the steepness of the slope of the table will also matter. This is like the electric field and the easiest way to get a measure of it is to find the voltage across the object conducting the electric current because field is measured in volts per metre.

A small current, say up to 0·5 A, is passed through a wire and the potential difference across the wire measured. Either circuit of Fig. 7.18 will do if the meters are good, but (*b*) is better if the voltmeter is less good. Using several cells in the battery the size of the current can be altered and each time the readings of both meters are taken. The figures can be put in the form of a graph and they typically lie on a good straight line passing through (0, 0) as sketched in Fig. 7.19. The conclusion is that current (I) and p.d. (V) are directly proportional, or putting it another way $\dfrac{V}{I}$ is the same for each setting, dependent on the characteristics of the wire. This relation is called Ohm's Law.

We generally write it as $V = IR$, where R is the wire's property called its **resistance** which fixes the value of the ratio $\dfrac{V}{I}$. Resistance is measured in **ohms**, symbol Ω, and a 1 Ω conductor allows a current of

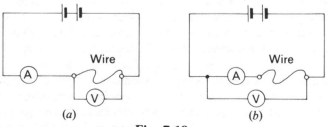

(a) (b)

Fig. 7.18

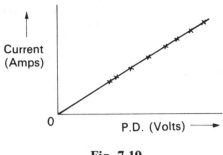

Fig. 7.19

1 A to pass when its ends are at a potential difference of 1 V. Resistors are in very common use in all kinds of electrical appliances and values range from $\frac{1}{100}$ Ω or less up to hundreds of millions of ohms.

7.22 Resistance and temperature

Coming back again to the pin-table idea (Fig. 7.16), it fails in one respect compared to a real solid material. The pins should be vibrating to simulate the motion of the atoms. What if they were? Fewer balls would then make the trip per second, that is there would be a higher resistance offered to the flow of balls. In electrical terms, the hotter the material (the more violent the vibration) the higher the resistance ought to be.

In general this turns out to be the case, and platinum resistance thermometers (Section 4.10) are devices which make use of the effect. An electric lamp with a tungsten filament takes a larger current when first switched on than when it has reached its operating temperature.

There are a growing number of *semiconductors* nowadays though which show a large *decrease* in resistance when temperature is increased, just the opposite change to the one expected from the above argument. The overriding mechanism here is that the increased temperature makes more electrons available for conduction purposes, which overwhelms the normal rise in resistance caused by the extra agitation of the atoms. (Semiconductors are materials whose electrical resistance lies between those of the good conductors like copper and aluminium and silver, and those of the insulating materials like polythene and PVC and rubber. The chief substances involved are germanium, silicon, arsenic, selenium and gallium, and one of the greatest technological advances of recent years has been the development of these materials into diodes, transistors, thyristors and all the sophisticated devices of modern electronics – see Chapter 8.)

Another odd thing worth mentioning is that at very low temperatures most metals become *superconducting*, that is to say they lose all electrical resistance. An electric current once started in such a condition would flow indefinitely without requiring further energy to maintain it. This has an important application in computers and possibly could be used in a very efficient electric motor which one day might power small cars.

7.23 Resistors in series

If three resistors of resistances R_1, R_2 and R_3 are connected end-to-end in **series** connection (Fig. 7.20) they could be replaced by a single resistor R whose value is given by:

$$R = R_1 + R_2 + R_3$$

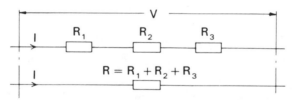

Fig. 7.20

To prove this we write down the potential difference across each resistor, add them up and equate the sum to the total potential difference:

$$IR_1 + IR_2 + IR_3 = IR = V$$

Thus resistors of 4 Ω, 6 Ω and 12 Ω connected in series are equivalent to a single resistor of value 22 Ω.

7.24 Resistors in parallel

If the resistors are in parallel connection as in Fig. 7.21, the combined

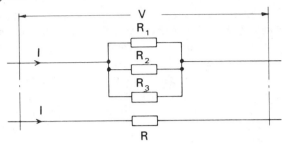

Fig. 7.21

effect is the same as that of a single resistor R whose value is given by

$$\frac{1}{R} = \frac{1}{R_1} + \frac{1}{R_2} + \frac{1}{R_3}$$

This result is obtained by splitting the current I into the parts I_1, I_2, I_3 which flow each way.

Using

$$I = I_1 + I_2 + I_3$$

we can write

$$\frac{V}{R} = \frac{V}{R_1} + \frac{V}{R_2} + \frac{V}{R_3}$$

and then divide by V.

In this connection, the $4\,\Omega$, $6\,\Omega$ and $12\,\Omega$ resistors could be replaced by a single one of $2\,\Omega$.

7.25 Using Ohm's Law – design of a multimeter

A meter would be bought by specifying two factors – its resistance and the current for full-scale deflection; e.g. $20\,\Omega$, 4 mA f.s.d. Such a meter can be adapted to read any current range in excess of 4 mA and any p.d. above 80 mV.

Consider the problem of coping with a current of 0·4 A, or 400 mA. Of this current, only 4 mA can be allowed to flow through the meter itself or the movement of the instrument will go beyond the limit of its scale and may be damaged. The other 396 mA must be shunted around the meter by an alternative path – Fig. 7.22. How large must R

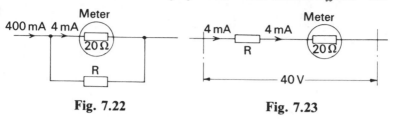

Fig. 7.22   **Fig. 7.23**

be so that this ratio of 396 : 4 is exact? The 20 Ω and R must of course be in the same ratio.

$$\frac{R}{20} = \frac{4}{396}$$

giving $$R = \frac{80}{396}\,\Omega$$

$$= \underline{0.202\ \Omega}$$

A similar calculation will be needed for each current range.

To deal with a large voltage, say 40 V, we must arrange that this only drives the permitted maximum of 4 mA through the meter by putting a resistor in series with it – Fig. 7.23. Using Ohm's Law here,

$$40 = \frac{4}{1000}\,(R + 20)$$

so $$R = \underline{9980\ \Omega}$$

Fig. 7.24 shows how a basic meter can be adapted in a simple way to deal with several ranges of current and voltage. In this design different terminals are needed for the amps and volts ranges (the reader can try to sketch the necessary connections using only one pair of terminals). A multi-way switch is needed and the diagram shows three current ranges and three voltage ranges with two switch positions connected directly to the meter.

7.26 Measuring resistance – the ohm-meter

Many multimeters include a resistance range as well as current and voltage. This can be very useful for checking the continuity of cables and wires and fuses. A small battery is incorporated in the case

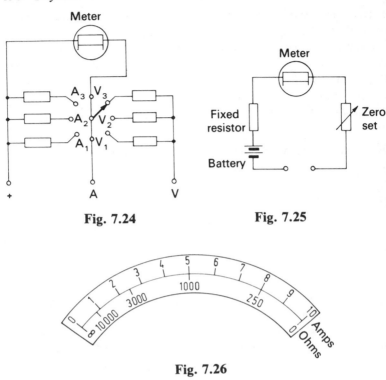

Fig. 7.24   Fig. 7.25

Fig. 7.26

(Fig. 7.25) together with a resistor in series with the meter. When the terminals are short-circuited the deflection is arranged to be full scale by a suitable choice of the resistor. If another resistor is connected between the terminals the reading will be less than full scale. This means that the zero of the ohms range must be at the right-hand side of the scale and an infinitely high resistor would leave the needle undeflected at the left-hand side. The resistance scale is not evenly divided and Fig. 7.26 shows a typically marked scale. The ohm-meter often includes a 'zero-set' adjustable resistor to allow for the battery's running down.

7.27 Electric currents make things hot

Tungsten filament lamps, electric fires, immersion heaters, electric blankets, etc. are commonplace applications of the heating effect of an electric current. It is easy to work out how much heat is generated per second.

Suppose a current of I amps flows through a wire across which there is a potential difference of V volts – Fig. 7.27. I coulombs flow through the wire every second and each coulomb requires V joules of energy to get through (Section 7.5). Thus IV joules per second are involved in the process and this energy is transformed into heat at that rate.

Heat produced per second = IV watts

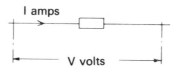

Fig. 7.27

If we are using a wire which follows Ohm's Law we can express the rate of energy transfer in terms of its resistance in two ways:

$$IV = I \times IR = I^2 R$$

$$\text{or} \qquad IV = \frac{V}{R} \times V = \frac{V^2}{R}.$$

These last two expressions depend on the meaning we give to resistance, but the product IV is always applicable since it follows from the definition of voltage.

7.28 Beware Ohm's Law!

Such a simple connection between p.d. and current cannot really be expected to describe the many different types of electrical components used nowadays nor even the behaviour of the simplest ones over their full ranges. In fact a too trusting belief in Ohm's Law can lead to unfortunate errors. Very few things obey it completely. The test is whether the values of p.d. and current lie on a straight line *through* (0, 0).

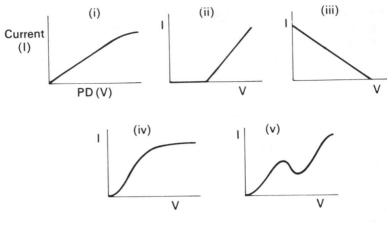

Fig. 7.28

Some typical graphs for different devices are shown in Fig. 7.28.

(i) A wire, getting hot at high currents.
(ii) Flow through a dilute solution of sulphuric acid.
(iii) Current-voltage for a cell with internal resistance.
(iv) Curve for a thermionic diode.
(v) Curve for a thermionic tetrode or tunnel-effect diode.

Only the first of these obeys the law and that for only the low current region.

7.29 Internal resistance

If you connect a wire from $+$ to $-$ of a car battery the wire will quickly get hot enough to burn off its insulation and possibly melt itself. The same thing done to a torch battery produces no such fireworks. Why not? Where is the resistor which limits the current in the second case?

The only possible answer is inside the battery itself, and it is called **internal resistance**. The size of it depends on the construction of the cells of the battery. Lead–acid accumulators have a very low internal resistance, perhaps $\frac{1}{100}$ Ω, hence the high current when they are short circuited. Dry cells as used in torches and for transistorised radios may have a value as high as 15 Ω.

One advantage of a high internal resistance is that there is no chance of damage by accidental short circuits, but such batteries cannot ever deliver a high current. Their use is restricted to devices needing only fractions of an amp. Accumulators, on the other hand, can deliver hundreds of amps (as when starting a car), but short circuits can be dangerous.

7.30 Ohm's Law for a full circuit

Consider a single cell such as a U2 torch cell; it has internal resistance $r \, \Omega$ and the chemistry of what is inside it determines the size of the electromotive force or e.m.f. (Section 7.4) which, say, is E volts. When this cell is connected to a resistor of $R \, \Omega$ (the bulb of the torch) the current which flows, I, will be decided by E volts and the two resistors R and r in series – Fig. 7.29.

$$I = \frac{E}{R + r}$$

or
$$E = I(R + r)$$
$$= IR + Ir$$

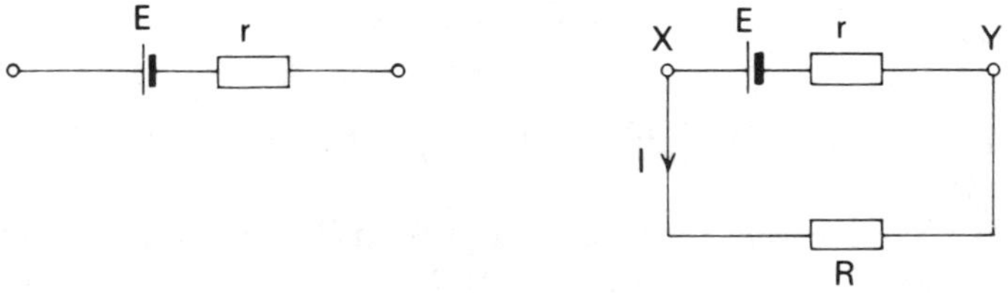

Fig. 7.29

As far as the external resistor R is concerned it 'sees' only the two terminals XY of the cell and a p.d. of V volts between them. Using Ohm's Law for the resistor R, we know $V = IR$. So putting V for IR in the last equation,

$$E = V + Ir$$

or
$$\boxed{V = E - Ir}$$

This means that the available terminal p.d. (V) is less than that which the cell develops (E) by the amount needed to drive the current through the cell itself (Ir). V and E will only be equal if either I or r is zero. For a torch cell, for instance, values might be $E = 1·5$ V; $r = 4\Omega$. If this cell delivers $\frac{1}{10}$ A, the available p.d. would be only

$$V = 1·5 - \tfrac{1}{10} \times 4$$
$$= \underline{1·1V}$$

7.31 The electricity bill – what we pay for (see also Section 3.9)

The thing we want from the electricity board is *energy*. This is what keeps us warm, gives us light and enables our electric machinery to function. We pay for joules. The 'electricity' meter is really a joulemeter and one day may perhaps be marked in joules, or more likely megajoules.

On the electricity bill the number of 'units' consumed per quarter is shown and the amount we owe worked out at whatever rate is operating at the time. How big is a unit? Another name for it is a kilowatt-hour.

$$1 \text{ kWatt-hour} = 3\,600\,000 \text{ joules}$$
$$= 3·6 \text{ Megajoules.}$$

We buy electrical energy in $3·6$ MJ lumps, the 'units' of the electricity boards.

An electric fire may be rated at $1\frac{1}{2}$ kW. If it is left on for 2 hours it consumes 3 kW h of energy, i.e. $10·8$ MJ.

7.32 Sending electrical energy from power station to user – direct or alternating current?

Suppose an industrial concern requires 1 MW of electrical power; that is, 1 000 000 joules per second. The result of Section 7.27 shows that the product of potential difference and current is what matters; we could deliver the power at 100 000 volts and 10 amps or 10 000 volts and 100 amps. Which is preferable?

The energy must get from the power station to the factory by means of cables – suppose these have an overall resistance of 5 Ω. The power station A which generates at 100 000 V sends 10 A through these

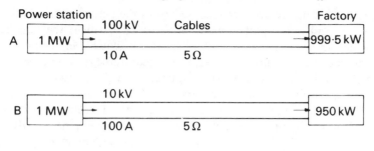

Fig. 7.30

cables. 10 A through 5 Ω produces 500 W of heat energy, so the factory can only receive 999 500 watts instead of 1 000 000 – Fig. 7.30.

Power station *B* generates at 10 000 V and sends 100 A through the cables. The heat generated this time is 50 000 W, so only 950 000 watts find their way to the factory. Clearly the high voltage/low current method wastes less energy on the way and is the one to aim for, but it is not quite such a simple thing to decide. High voltages around the workshop or home are dangerous, so while we want to *transmit* the power at high voltage we wish to *use* it at a lower voltage. What we need is a device which will change voltages up and down without, if possible, wasting any power in the process.

This decision of economics leads us to choose AC for power transmission, for while there *are* DC converters their efficiencies are rather low. It is the 93 % efficiency of transformers which forces us to decide on AC. In Section 7.49 these devices are discussed in more detail, but the point here is that electrical power transmission is cheaper at high voltages and using alternating current we can swap voltages with only modest loss of power – so power is transmitted using AC.

7.33 Electromagnets

Take a 6 or 7 cm iron nail, wrap a few tens of turns of insulated copper wire around it and connect the ends of the wire to a battery. The result is an **electromagnet**. The nail will now be able to pick up small panel pins, perhaps quite a lot; it certainly could not do so before the battery was connected. What has happened to it?

Clearly it has become magnetised by some action of the electric current, and if the nail is made of fairly soft iron it will lose its magnetism when the circuit is broken. It is a switchable magnet and finds ready use in a thousand and one devices – tape recorders, electric bells and buzzers, relays, scrap metal yards, automatic sorting machines, cyclotrons, loudspeakers, transformers, etc.

The main factors governing the strength of an electromagnet can be found easily using the nail and some small panel pins. A battery capable of giving several amps and enough copper wire to wrap many turns round the nail are needed, together with an ammeter and rheostat to vary the current – Fig. 7.31. For a fixed number of turns, say 50, the current can be increased step by step and the strength of the nail magnet judged by how many panel pins it will hold from a dish of pins gradually lowered from one end. The pins must either be weighed or counted.

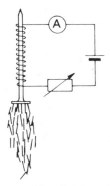

Fig. 7.31

A similar technique with a fixed current of say 2 A and different numbers of turns of wire from 10 to 500 will show how the strength depends on this factor. Typical curves are shown in Fig. 7.32; both are of the same shape and both flatten off, showing that there is a limit to the magnetisability of a piece of iron or steel. When this happens the nail is said to be **saturated** and is magnetised as powerfully as possible. A larger nail has similar curves, but more pins are attracted at each stage and its saturation strength is greater. Different kinds of steel show generally the same behaviour, some being more easily magnetised than others, some more powerfully than others.

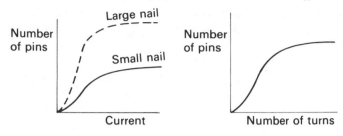

Fig. 7.32

7.34 Permanent and temporary effects

There is one striking difference between almost pure iron and various steels which matters greatly when applications are devised to use the electromagnetic effect. Pure or soft iron (containing about 0·1 % carbon) will drop almost all its load when the current is switched off whereas harder steels (perhaps up to 0·8 % carbon) retain their magnetism even when the cause of their magnetisation is removed. Permanent magnets obviously have to be made from steel or alloys of similar properties and devices which require 'switch-off-able' magnets will have to use soft iron for their construction.

7.35 Magnetic field patterns in a coil

To understand what happens to a piece of iron placed inside a coil of wire we must extend the idea of 'field' used in Section 7.11 to explain forces felt by charges of electricity. (Indeed the concept of a 'field of force' is a useful one for any phenomenon where effects are felt at a distance from the source of disturbance and has found expression in magnetism, electricity, gravitation and nuclear physics.)

Small magnetic particles are needed to reveal magnetic field patterns and iron filings are very suitable. To find the field pattern inside a coil it must be made large enough to sprinkle filings on a card down the length of the coil. A pepper pot can be used to sprinkle the filings and several amps will be needed if a convincing pattern is to be seen – Fig. 7.33. The card may have to be tapped lightly. The field lines run through the length of the coil.

Which way do lines go? This is another chance to make a choice and it was done many years ago. If small magnetic compasses are placed

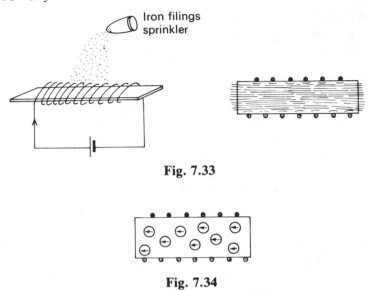

Fig. 7.33

Fig. 7.34

on the card in place of the filings the way they point is taken to be the way the field lines go. With current coming up at the bottom edge of Fig. 7.34, over and down at the top edge, the compasses point towards the left.

7.36 Magnetic field pattern near a wire

If a similar experiment is done to find the field pattern near to a single wire carrying an electric current the result is sketched in Fig. 7.35.

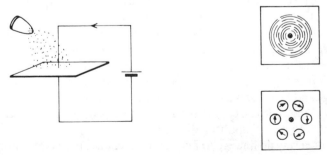

Fig. 7.35

Many amps are needed to make filings show a marked set of lines or, alternatively, many strands of wire tied closely together each carrying the same current.

7.37 A theory of magnetism

Why should a piece of iron placed in a field like Fig. 7.34 become magnetised? The full answer to this is long and difficult, but a single idea is worth looking at. It is thought that inside iron there are rather large regions (large on the atomic scale) which are already magnetised by some internal mechanism in the iron. These regions, called **domains**, are determined partly by crystal boundaries within the lattice of molecular iron, and in the unmagnetised state have their directions of magnetism all randomly distributed, so no preferred direction can be detected for a complete chunk of iron.

What magnetisation does is to cause changes in the domain boundaries so that those which are in the same direction as the magnetising field of the coil grow at the expense of those directed oppositely. The result is a gradual alignment of the domain directions which gives a chunk of iron a definite direction of magnetisation.

This idea explains many simple observations about magnets.

(i) When all domain directions are aligned the iron will be saturated.
(ii) If you break a magnetised piece in half you will be left with two magnets.
(iii) Causing physical shock to a magnet can cause it to lose some magnetisation as domain boundaries revert back to the random state.
(iv) Heating a magnet will destroy domain boundaries due to vibration of the iron atoms and so demagnetise the material.

7.38 Magnetic field patterns near magnets

Field patterns are sketched in Fig.7.36 for some simple arrangements of magnets. By 'N' and 'S' is meant the end of the magnet which points roughly North or South when the magnet is held by a fine thread. Like electric field lines the closeness of the lines to each other is a measure of the strength of a field. From these sketches we can see that powerful fields are found generally between two poles of opposite kind and the field lines are almost straight in the middle region.

Magnets can be obtained now with N and S *faces* rather than N and

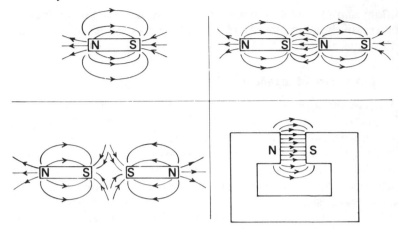

Fig. 7.36

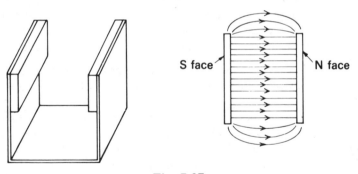

Fig. 7.37

S ends. Two of these placed a few cm apart give a fairly even and uniform field – Fig. 7.37. The best way to hold them is on a soft iron ⌊_⌋-shaped yoke.

7.39 Interaction between two fields – catapult effect

Suppose a wire carrying a current is placed in a straight-across field like that of Fig. 7.33. We will have two fields superimposed now and the result will be the sum of the two – but how to add up fields? Well, fields are only forces felt by certain test objects, so they will add up like forces did in Section 2.1. Where the two fields reinforce each other will be an increased field, where they oppose will be a weakened field. Fig. 7.38 shows what happens to the field pattern for a wire carrying a current up out of the page; there is a very unbalanced field, much stronger on one side of the wire than the other and perhaps not surprisingly the wire experiences a force towards the weaker field region. This force is at right angles to both the straight-across field and the current directions, rather like a catapult.

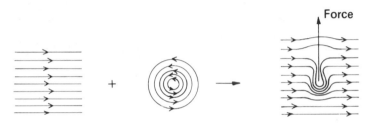

Fig. 7.38

7.40 A simple motor

The catapult effect can be used to make a simple motor. A matchbox or wooden block is used to wind a coil of insulated wire on and the ends of the coil touch two stiff upright contacts – Fig. 7.39. The coil can be supported by a large pin or knitting needle, insulated with tape and passing through simple bearings like the heads of split pins. The whole thing mounted on a block of wood can be fitted between two magnets, and a cell connected to the stiff contacts. While the contacts and the ends of the coil are touching each other, current flows round the coil and one side is catapulted up, the other down with sufficient force for the coil to make half a revolution, remake contact and be given a further push round. Fig. 7.40 shows the directions of the forces on the sides of the coil.

Brush-contact motors of commercial design follow this basic principle, but usually contain several coils wound on a laminated iron former and the magnet has curved pole-pieces. The magnetic field is generally provided by an electromagnet rather than permanent magnets.

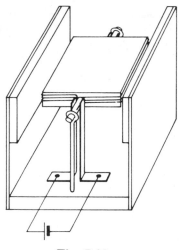

Fig. 7.39

Fig. 7.40

7.41 A simple meter

If the coil used to make the motor in the last section is prevented from going all the way round by a simple spring system it becomes an ammeter – the actual angle turned through depends on the size of the current and the strength of the springs. The wire itself can be wound into a rough spiral at each end of the coil and a light straw attached to the coil will indicate the amount of deflection – Fig. 7.41.

7.42 Moving-coil meters

The construction of ammeters and voltmeters is basically the same as that outlined in the last section, but care is taken to make them more robust, more sensitive and designed so that equal steps of current cause equal changes of deflection. Fig. 7.42 shows a common construction.

7.43 Generating electricity

Usually in physics, if an effect is possible the reverse effect will also be possible. For instance, if a piece of iron is heated it expands; conversely, if a piece of iron is stretched it cools. Again, oxygen and

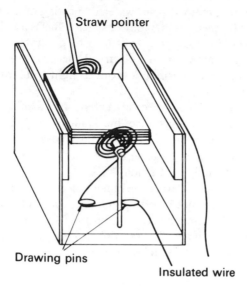

Straw pointer

Drawing pins

Insulated wire

Fig. 7.41

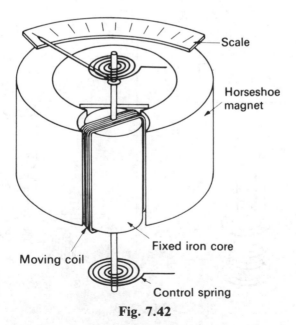

Scale

Horseshoe magnet

Fixed iron core

Moving coil

Control spring

Fig. 7.42

hydrogen can be obtained from water by passing an electric current through it; the fuel cells used to power space vehicles use the reverse of this and combine oxygen and hydrogen to generate electricity and water. An electric lamp converts electrical energy into heat and light; photoelectric cells and thermopiles convert light and heat energy into electrical energy.

The combination of an electric current and a magnetic field as described in Section 7.40 produces mechanical energy in the coil of a motor. What is the reverse of this? From mechanical energy and a magnetic field we require to generate electrical energy. The effect was discovered accidentally by Faraday in 1831.

Whenever there is relative movement between a magnetic field pattern, however produced, and a conductor, an e.m.f. is developed in that conductor. The e.m.f. can cause a current to flow if there is a closed circuit in which it can act. Several different arrangements for giving the necessary change are sketched in Fig. 7.43. Perhaps the easiest to

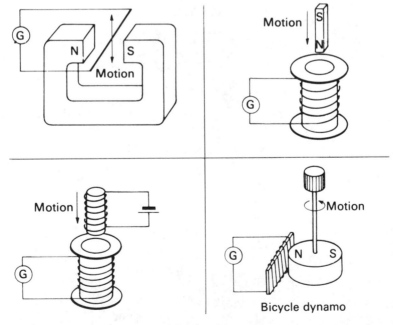

Fig. 7.43

use in practice is the ready-made motor of Section 7.40. If the cell is replaced by a galvanometer and the coil spun by hand there is a deflection of the meter as long as the coil is rotating. The faster the rotation the bigger the effect. Other ways of making a larger induced emf are to increase the number of turns on the coil, wrap them on a soft iron armature, use a coil of larger area and increase the strength of the magnet. These of course are just the factors which would make the motor more powerful in Section 7.40.

7.44 The law of cussedness

'If things can be awkward, they will be.' Why not have a motor connected to drive a dynamo which generates electricity to supply the motor which turns the dynamo which generates . . . etc.? Unfortunately this dream is unattainable due to the *direction* of the induced e.m.f., which is always such that if a current flows there will be forces acting to slow down the motion causing the original e.m.f.

Think of the simple motor of Fig. 7.40. Sending a current through its coil causes the coil to move in the magnetic field. As soon as this movement starts we have just the right condition for an induced e.m.f. in the coil. If this e.m.f. *helped* the flow of current the coil would move faster, which would mean a larger induced effect giving an even faster motor speed and so on, making a self-perpetuating system only needing to be started off to run on its own resources for ever! This clearly does not happen in real life and the reason is that the induced emf *hinders* the flow of current through the coil and acts as an automatic brake.

A neat illustration of the braking effect can be seen if a coin is suspended from a twisted thread. Ordinarily the thread uncoils rapidly and turns the coin quite quickly. If the same thing is done with the coin placed between the poles of a strong magnet (Fig. 7.44) a very much slower rate of spinning is seen. The current induced in the coin as it spins is always just in the right direction to slow it down.

7.45 Generators and dynamos – AC and DC

The simple electric motor of Section 7.40 will run backwards as a dynamo, as mentioned in Section 7.43, but if the motor is turned slowly a meter connected to the contacts does not show a steady reading. It goes in surges from zero up to a maximum, down to zero

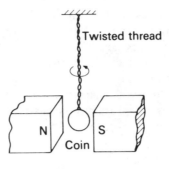

Fig. 7.44

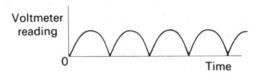

Fig. 7.45

again and so on. The voltage varies with time like the sketch graph of Fig. 7.45 shows. Although this would be called a 'direct' voltage it clearly is not the same all the time, but if the speed of rotation is increased the meter does settle down in some 'average' level due no doubt to the fact that the coil system of the voltmeter has too much inertia to follow the rapid changes. This type of dynamo generates a varying voltage, but always in the same direction.

If the coil of the dynamo is altered so that, instead of the ends coming out to two terminals at one end which touch each side of the coil in turn, they emerge at opposite ends, contact can be maintained with each end all the time – a ring of aluminium foil round the axle will allow this to be done – Fig. 7.46. This time slow rotation causes the needle of the voltmeter to rise to a maximum in one direction, fall to zero, then go an equal amount in the opposite direction before returning to zero and repeating the procedure. A graph this time is like Fig. 7.47. For a fast rate of spinning the meter registers zero.

This second voltage is called **alternating** since it acts in turn – and equally – in the forward and reverse directions, and the device might

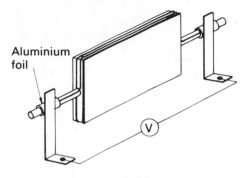

Fig. 7.46

Fig. 7.47

be called an **alternator**. A simple bicycle dynamo like Fig. 7.43 generates an alternating voltage and shows the advantage of keeping the coil fixed instead of the magnetic field (remember only *relative* motion is needed). The problem of moving contacts is overcome and many coils can be used. Real alternators used in power stations work in this way.

7.46 How to give a number to alternating currents?

The average value of the graph sketched in Fig. 7.48 is zero. Any moving coil meter placed in a circuit carrying AC will read zero or else dither slightly to each side of it. How can we give a number-value to an alternating current or voltage?

There is one thing which electric currents do which does not depend on the direction of flow – they cause a conductor to get hot and reversing the current does not cool it down! We use this effect to label what is meant by, say, '5 amp AC' in the following way:

If 5 amp DC causes W joules of energy to be converted into heat every second in a conductor, the value of an alternating current which produces the same rate of heating of W joules per second in the same

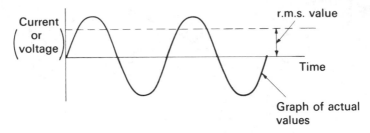

Fig. 7.48

conductor is also called 5 amp. This means that the value of 5 A must be some sort of average value of the current, though obviously not the simple average since that is zero. It is called the **root mean square** value, or r.m.s. value, and is about 71 % of the maximum current for an ordinary AC, like Fig. 7.48. To quote '5 A AC' means a current which fluctuates between about $+7$ and -7 A. '240 V AC' means a voltage swinging between $+340$ and -340 V.

7.47 Induction – electromagnetic link-up

Referring back to Section 7.43 the only thing required for an e.m.f. to be developed in a coil or wire was to have a changing magnetic field around it. In Fig. 7.49 it is enough to switch on or off in the main circuit for the galvanometer to flick, but once the current *is* on or off there is no deflection. If a lump of soft iron is used to 'join' the two coils, as in Fig. 7.50, the effect is much greater, but still only present as the switch is opened or closed. A complete magnetic 'circuit' of soft iron gives an even larger effect – Fig. 7.51. The flicks of the galva-

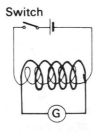

Fig. 7.49

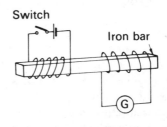

Fig. 7.50

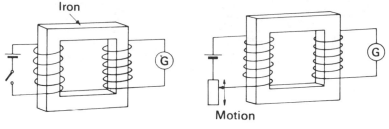

Fig. 7.51   **Fig. 7.52**

nometer also occur when a rheostat included in the cell circuit has its slider moved to change the size of the current flowing – Fig. 7.52. All these simple alterations are ways of causing a *change* in the magnetic field lines in the coil connected to the galvanometer.

Any method of giving regular changes would 'induce' regular deflections and there are two obvious ways of doing this – to arrange for *mechanical* or *electrical* switching, or reversal of the main circuit current.

7.48 Method 1 – the induction coil

The sparking plugs in a car are fed with a very high voltage which causes the spark to ignite the petrol–air mixture in the cylinders. This high voltage has to be derived from the 12 V car battery or dynamo and the arrangement used to achieve it is shown in Fig. 7.53. The rotation of the engine opens the contacts which break the circuit and stops the current flowing in the coil. The rapid collapse of the magnetic field caused by switching off induces a very high emf in a second coil having many turns of wire, and the emf acts across the electrodes of the sparking plugs. This occurs four times in one revolution of the rotor arm for a 4-cylinder engine sending a pulse of high voltage to each of the four plugs.

7.49 Method 2 – the transformer

If alternating current is sent through the turns of the first coil in Fig. 7.52, since it is continuously changing in value from, say, 2 amps in one direction to 2 amps in the reverse direction, there will always be a changing field and therefore always an induced emf in the second

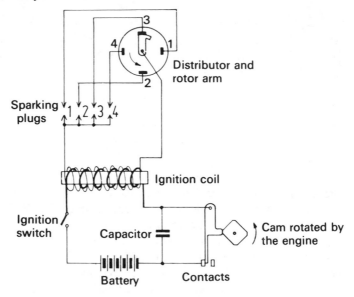

Fig. 7.53

coil. This will also alter in size and direction like the changing current, but will always oppose the effect causing it (Section 7.44). We have an *alternating* voltage induced in the second coil.

This can be large enough to light a bulb or to work a television set depending simply on the number of turns of wire on the two coils. If the **transformer**, as it is called, were perfect we could work out that:

$$\frac{\text{Alternating voltage applied to coil 1}}{\text{Alternating voltage induced in coil 2}} = \frac{\text{No. of turns on coil 1}}{\text{No. of turns on coil 2}}$$

Fig. 7.54 shows two simple arrangements which enable alternating voltages to be 'stepped up' or 'stepped down'.

In Section 7.32 it was realised that a machine capable of changing voltages up or down would be invaluable for high voltage transmission of electrical power, and it is the transformer which makes AC the choice for transmitting electricity rather than DC, for which there is no comparably efficient machine available. One simple design feature which makes the transformer so efficient is shown in Fig. 7.55; both

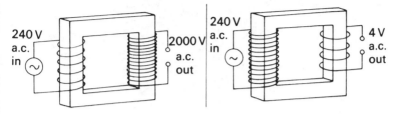

Fig. 7.54

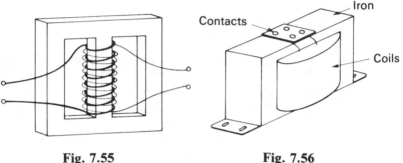

Fig. 7.55   **Fig. 7.56**

coils are wound (insulated from each other, of course) on the same piece of soft iron to make the electromagnetic link-up as intimate as possible. The soft iron is also made in layers separated from each other by thin coats of insulating material. Fig. 7.56 shows a real transformer with the contacts brought out to the top of the laminated soft iron core.

7.50 Transformers do not give something for nothing

The prospect of transforming alternating voltages up and down tempts us to think of the transformer as an amplifier or multiplier able to give out more than it takes in. Many attempts to make such a device have been made in the past, but all have fallen foul of one or other scientific law and the transformer is certainly no exception.

Transformers *can* multiply voltage, but what they cannot do is to give us more *power*. Since power is worked out by multiplying voltage by current (Section 7.27) this must mean that any increase of voltage is

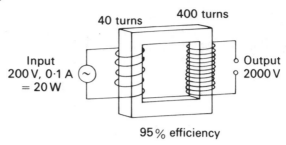

40 turns   400 turns

Input
200 V, 0·1 A
= 20 W

Output
2000 V

95 % efficiency

Fig. 7.57

accompanied by a proportionate fall in current. In fact it is not as simple as this because no transformer is 100 % efficient, so there is a greater loss of current than we expect. Here is an example worked in figures to illustrate the point, using Fig. 7.57.

If the transformer is 95 % efficient, 20 watts input power will yield 19 watts output power. If the turns ratio is 40:400 we will get a voltage jump of 10 times, so 200 volts input gives 2000 volts output, but with only $\frac{19}{2000}$ amp of current, i.e. 9·5 mA instead of the 'ideal' 10 mA. Such a transformer can deliver only 9·5 mA, even if the output coil is short-circuited.

Calculations like this also show the importance of considering the thickness of the wires in the two coils as well as their numbers of turns. In this case the primary coil will have to be capable of handling 100 mA and must be thick enough for there to be negligible power loss in it when that current flows. The secondary coil though can be made of much thinner wire since it will have to conduct only 9·5 mA.

7.51 Changing AC to DC

Almost all electronic devices operate at certain definite voltages or currents or vary only slightly from these steady values, so although it is convenient to generate and transmit electrical energy in the form of alternating current we have the task of converting it into direct current for most appliances such as television sets, radios, record players, tape recorders, etc.

Just as the valves in the veins or heart inside our body allow the blood to flow only one way, so we need an electrical device which does the same thing for the flow of current. There are in fact many different types of component which will do this job in an electrical circuit, called

collectively **diodes** and denoted by the circuit symbol.

The 'arrow' indicates the direction in which conventional currents are able to flow. We will not discuss the different types of diodes in detail here (they range from metal oxide rectifiers, through point contact 'cat's whiskers', to silicon or germanium semiconductor diodes, mercury arc rectifiers and thermionic diode valves), but concentrate on the job they do. A useful tool to have here is the Cathode Ray Oscilloscope which will be dealt with fully in the next chapter, Section 9.8. The important property this device has is that it can draw for us a voltage/time graph and display what is happening at various points in a circuit.

In Fig. 7.58 an alternating supply sends a current through a resistor. Looking across the points *AA* the oscilloscope shows a voltage/time curve like the one sketched alongside the circuit. If a diode is inserted (Fig. 7.59) the curve changes showing that the alternating voltage has been **rectified** to act always in one direction, though not at a steady value.

This simple rectifier circuit 'wastes' half the available voltage by simply not conducting for half the time. If the 'wrong' half of the curve could be turned upside down and added to the half already being

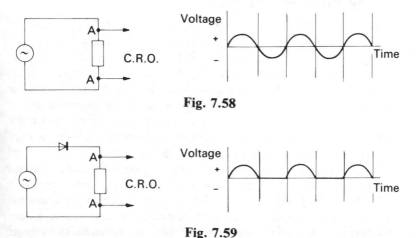

Fig. 7.58

Fig. 7.59

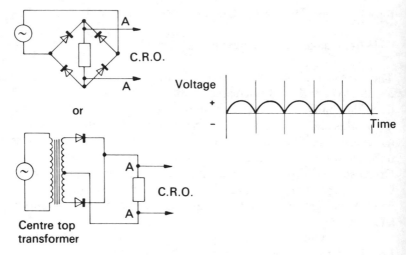

Fig. 7.60

conducted there would be a current flowing all the time in the same direction. Fig. 7.60 shows two ways of doing this and the reader is left to work out why both circuits give an output graph like the one drawn for them.

7.52 Smoothing out the ripples

The rectified voltage of Fig. 7.60 is all in one direction but does not remain steady. For things like electric motors in model racing cars or electric trains or battery chargers this does not matter, but for a supply to drive an amplifier it will not do. The effect of using a crudely rectified voltage for an amplifier would be disastrous since the strong alternating component of this supply (something alters for every hump of the curve) would itself be amplified and give rise to a loud humming or buzzing noise. We need a 'smooth' voltage graph with little or no ripple for applications like this.

Advantage can be taken of the way a capacitor stores charge (Section 7.7). Fig. 7.61 shows the output points *AA* connected to a capacitor of large value. As the voltage rises the capacitor becomes charged up during the first hump of the curve and subsequently discharges slowly when the voltage falls. The next rise of voltage

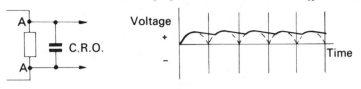

Fig. 7.61

charges the capacitor again so that after the initial rise there is very little variation in the voltage across it. The capacitor is acting like a large reservoir of charge which is topped up every time the voltage rises.

An even more efficient smoothing unit is shown in Fig. 7.62 using two large capacitors and an iron-cored coil or 'choke'.

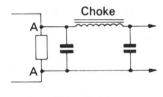

Fig. 7.62

8

Elements of Electronics

8.1 A growth industry

The development of modern electronics can be traced back to the discovery of 'semiconducting' materials and the invention of the transistor in the late 1940s. Since then the ability to produce large quantities of the substances which make up these components, and the skill in designing smaller and smaller circuitry in which to use them, has resulted in a real electronics revolution. The rate of progress since 1950 has been accelerating steadily so that it is now possible for new ideas to be conceived, developed, tested and put into production in such a short time that few people are able to keep up to date except in very specialised areas.

It is easy to take electronic devices for granted now that they have become such common features of industry, entertainment, commerce, education and leisure. Transistor radios, electronic games, home computers, video recorders, automatic cash registers, robot-operated manufacturing processes, electronic timers, digital clocks, calculators, electronic organs, synthesisers, lasers, infra-red cameras, cardiac pacemakers, etc., represent just some of the vast range of appliances in which electronics is put to practical use. This chapter introduces the basic principles of many electronically operated devices.

8.2 Semiconductors: n-type and p-type

Most metals, especially copper and silver, conduct electricity very easily, whilst plastic materials like Perspex and PVC are used as insulators because they hardly conduct at all. Substances which behave electrically in an intermediate way between conductors and insulators are called **semiconductors**: their resistances are usually fairly high and variable. Germanium and silicon are examples of semiconducting elements whose resistivities are high but fall markedly if they are heated.

A more useful material can be obtained by adding very small quantities of other elements – such as phosphorus, arsenic or boron – to germanium or silicon in order to reduce their resistances. Two different types of semiconductors, called **n-type** and **p-type**, can be made in this way: for example, germanium with phosphorus as an impurity is n-type, and silicon with arsenic is p-type. The *n* and *p* refer to the *n*egative or *p*ositive charge carriers which conduct the electricity, negative carriers predominating in n-type semiconductors and positive in p-type.

If a current is made to flow through a piece of semiconducting material the electric charges are carried through it by both negative and positive carriers, rather like the way charges move through a conducting liquid. The carriers move in opposite directions according to the sign of their charges, there being many more negative ones in n-type materials and positive ones in p-type. Both cases actually involve electrons but the effects are as though there were two different kinds of carrier, one + and one −. Fig. 8.1 illustrates the passage of a current through n-type and p-type materials, with the carriers represented by open (+) and solid (−) circles. The former move in the direction of the conventional current and the latter against it, but the effect is the same for both, in that charges flow through the materials and conduction is effected.

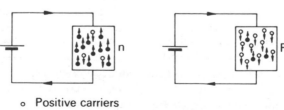

o Positive carriers

• Negative carriers

Fig. 8.1

8.3 The p-n junction

The working of many semiconductor devices depends on what happens when n-type and p-type materials are joined together. Without going into the mechanism too closely, the main effect is that conduction is very much easier in one direction than in the other – Fig. 8.2.

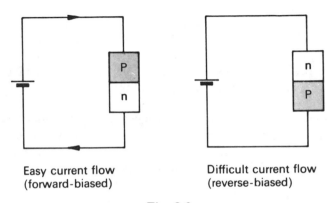

Easy current flow   Difficult current flow
(forward-biased)   (reverse-biased)

Fig. 8.2

The junction is said to be *forward-biased* if the p.d. across it is such that current can flow easily, and *reverse-biased* if the p.d. is such that very little current can flow. Thus a forward-biased junction will conduct easily and a reverse-biased one hardly at all.

The obvious use of such an effect is as a one-way device or diode of the type described in Section 7.51.

8.4 Diodes

A p-n junction between pieces of p- and n-type materials, together with two connecting wires and encased in a suitable container, forms a **diode** (Fig. 8.3). In each case the connection marked + is called the **anode** and that marked − the **cathode**. The conduction curves for a p-n junction diode (compare Fig. 7.28) are shown in Fig. 8.4 (note the different scales on the + and − axes). Such a diode would be described by its average forward current (e.g. 1 A) and its maximum reverse voltage (e.g. 50 V), neither of which should be exceeded under

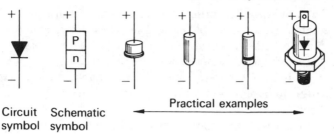

Circuit symbol   Schematic symbol   Practical examples

Fig. 8.3

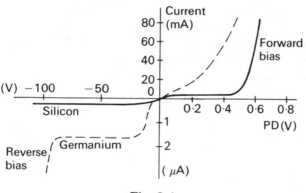

Fig. 8.4

normal conditions. Silicon diodes are usually preferred for rectifiers because of their higher values for each of these quantities.

Reference to books on electronics will show many other kinds of diodes designed for specific uses. A common one is the Zener diode, used as a stabiliser in power supplies under reverse voltage where at a certain value the current flowing does not affect the p.d. across the diode; this reference voltage (or Zener voltage) can be fixed in a range of values between, say, 3 V and 200 V.

8.5 Transistors

The basis of modern technology was the development of the **transistor** which is a three-element device – either a p-n-p combination or an n-p-n. The principal applications of transistors centre on their abilities to act as amplifiers and switches; that is, as *active* circuit components rather than *passive* ones like lamps, resistors or diodes. They are immediately suitable for control circuits, logic systems and radio equipment especially since it is possible for thousands of transistors to be mounted with other circuit elements on a small 'chip' of silicon (Section 8.15).

Compared with thermionic valves, which were used in electronic circuits previously, transistors represent a major advance. Apart from their much smaller sizes and greater reliability, transistors are easy to mass produce, they can operate at low voltages, they do not require separate heating circuits for energy supplies, they can deal with a wide range of currents and powers, and they can be put to a great number of uses. Miniaturisation in the manufacture of transistors has merely emphasised the overwhelming advantages of solid state devices over the large radio valves of the 1940s.

Single transistors come in various sizes and shapes with three connecting wires or points – Fig. 8.5.

Fig. 8.5

There are two basic types of transistor: the (ordinary) *junction transistor* and the *field effect transistor* (FET).

8.6 Junction transistors

Fig. 8.6 shows the two possible arrangements, p-n-p and n-p-n, and their circuit symbols. In each case a *base* of one type of semiconducting material is sandwiched between two pieces of the other type, called the *collector* and *emitter*. The n-p-n transistor is usually made of silicon and the p-n-p of germanium. The arrow indicates the direction of (conventional) current flow, under forward bias, between base and emitter.

Note that the actual construction of a transistor is much more

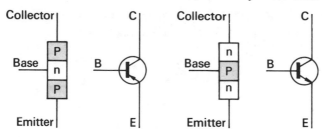

Fig. 8.6

complex than a simple double junction, the action depending upon the amount of impurity elements in each part and the thickness of the base.

For the transistor to work, the two junctions must be operated under different voltage conditions – the base-emitter junction forward-biased and the collector-base reverse-biased, as shown in Fig. 8.7 (compare with Fig. 8.2). (Although the figure shows a cell providing the necessary p.d. between base and emitter, in practice the voltage would be produced in other ways.)

The essential feature is that the small current flowing between base and emitter can control a much larger current between collector and emitter via the base, including switching it on and off. The sizes of typical currents are indicated on Fig. 8.8. (Note the resistor R, say 1 kΩ, which limits the current flowing into the base.)

If the base current (0·05 mA) were to be interrupted, the current into the collector could not pass through the transistor.

The junction transistor is a current-operated device in which one

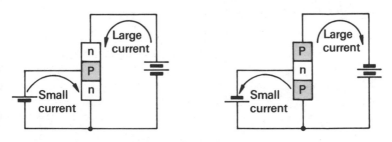

Fig. 8.7

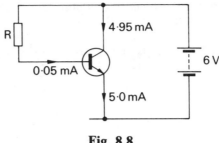

Fig. 8.8

current can switch on and off, or alter the value of, a much larger current, often between 10 and 1000 times its size depending on the particular transistor. Important variables are the size of each current, the current ratio and, in terms of safety, the maximum base current.

8.7 Field Effect Transistors (FET)

In this type of transistor, which also has three terminals, it is the voltage of the *gate* (G) which controls the current flowing between the *drain* (D) and the *source* (S). The behaviour of the FET depends on the way the n- and p-type semiconducting sections are constructed.

Fig. 8.9 shows the symbol for an FET and typical sizes of currents and voltages. Here, the essential feature is the change produced in the drain current for a given change in the gate voltage.

As a *voltage-operated* device, an FET is very useful for handling the output of a crystal pick-up of a record player, for example. A particular virtue of certain FETs is that they can very easily be

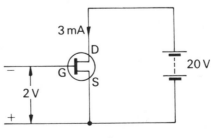

Fig. 8.9

manufactured in large numbers on a small silicon chip to form a single unit called an *integrated circuit* or microprocessor (Section 8.15).

Two different types of FET are in common use: the *junction-gate FET* (or JUGFET) and the *metal oxide semiconductor FET* (or MOSFET).

8.8 Transistor switching

The principle of the transistor's use as a switch can be seen from the circuit of Fig. 8.10.

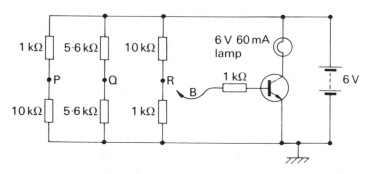

Fig. 8.10

It is only when the base lead B is connected to points P or Q that the lamp lights. Connecting B to point R (or leaving it unconnected) results in insufficient base current to allow the transistor to conduct between emitter and collector.

Note the two uses of resistors in this circuit. The 1 kΩ in the base lead is a *current-limiting* resistor to protect the transistor from too large a base current flowing. The pairs of resistors between the + and − conducting wires serve as *potential dividers*. These give at P, Q and R, respectively, voltages of 5·45, 3·0 and 0·54 V relative to the emitter.

The two circuits of Fig. 8.11 show how a simple light-operated alarm circuit can be made using a light-dependent resistor (LDR – see Section 8.11) which has a high resistance when dark but a much lower resistance in bright light.

The lamp in the first circuit would light in bright conditions, but in the second circuit in dark conditions (compare with the circuit of

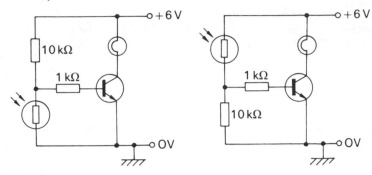

Fig. 8.11

Fig. 8.10). These circuits could be used to activate warning lamps if, say, an intruder switched room lights on (circuit 1) or if a lighting system were obstructed (circuit 2). The warning lamp could be placed some distance from the LDR, if necessary, and infra-red light beams could be used.

The same basic circuits could be used to detect fire or frost if the LDR were replaced by a thermistor (a semiconducting material whose resistance varies greatly with temperature – Section 8.11), and the lamp could in turn be replaced by a buzzer or a light emitting diode (LED). In the same way a relay could be inserted in place of the lamp to switch on a separate high current or high voltage circuit, as in Fig. 8.12.

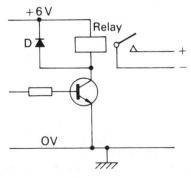

Fig. 8.12

A **relay** is an electromagnetically operated switch. In this case a diode D would be required to protect the transistor from the back e.m.f. generated by the relay coil when the latter is switched off by the transistor – compare Section 7.44.

8.9 Transistors and logic gates

A **logic gate** is a switching circuit which gives high or low outputs depending on the signals fed into it. Several different types of gate can be designed depending on the switching properties of the transistors used.

NOT gate   A gate whose output is high when its input is low (*not* high), or low when its input is high (*not* low).

In Fig. 8.13, if the input is high (say 5 V), the transistor will be switched 'on' and a large collector current will flow through the resistor R_L giving a p.d. across it which makes the output voltage low (i.e. very near O V). If the input is low (say 0·5 V), hardly any current flows through the collector, since the transistor is switched 'off', giving a small p.d. across R_L and a high output voltage (almost 6 V).

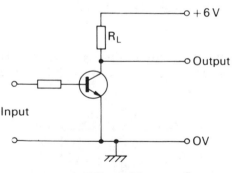

Fig. 8.13

The action can be summarised by a *truth table* in which 1 indicates high and 0 low, and the whole represented by a single circuit symbol:

Input	Output
0	1
1	0

NOR gate This gate has two (or more) inputs and its output is high only if neither one input *nor* the other is high (i.e. both inputs are low). The circuit, the truth table and the symbol are shown in Fig. 8.14. (From Fig. 8.13 it is clear that a NOT gate is a single input NOR gate.)

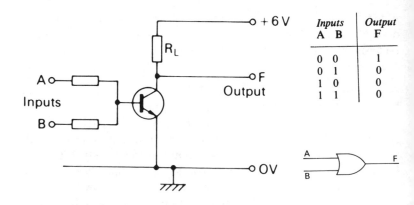

Fig. 8.14

OR gate A gate whose output is high if either one input *or* another is high (or if both are high). The circuit is a NOR gate followed by a NOT gate – Fig. 8.15.

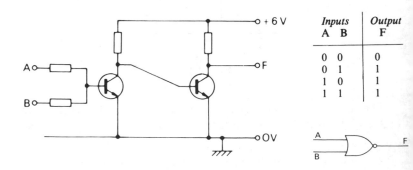

Fig. 8.15

NAND gate   A gate whose output F is *not* high (i.e. low) when both inputs A *and* B are high.

Inputs A B	Output F
0 0	1
0 1	1
1 0	1
1 1	0

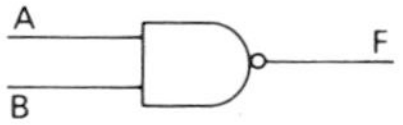

AND gate   A gate whose output F is high only when both inputs A *and* B are high. (This is a NAND gate followed by a NOT gate.)

Inputs A B	Output F
0 0	0
0 1	0
1 0	0
1 1	1

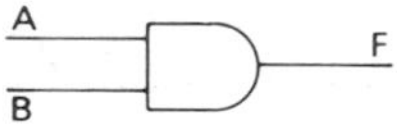

Exclusive OR gate   A gate whose output F is high only if either one, but not both, of the two inputs is high (i.e. only if the two inputs are different).

Inputs A B	Output F
0 0	0
0 1	1
1 0	1
1 1	0

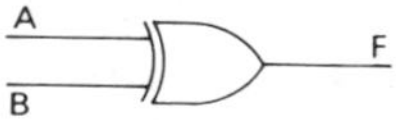

Circuits have been shown here only for the simplest gates, and it is not always a matter of just adding on component circuits to make a more complex one, but the truth table approach is useful in deciding what a sequence of gates would actually do in practice. For example, if the set of NAND gates in Fig. 8.16 is tackled by a series of truth tables, it will be found to be an exclusive-OR gate.

In practice all the separate NAND gates could be built into a single integrated circuit (IC – see Section 8.15) with the facility to interconnect them in a variety of ways.

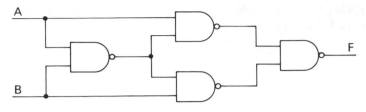

Fig. 8.16

The simple ideas described above form the basis of quite complicated circuits which fall beyond the scope of this book. Suffice it to say that combinations of logic gates are used in many digital electronic devices, such as computers, decimal converters, decision-making circuits, encoders and decoders, adders, multiplexers, memory circuits and many more.

8.10 Transistor amplifiers

In Section 8.6 the action of transistors in operation was seen to lead to two main uses – as switches and as amplifiers. The latter use was based on the control of a large collector current by a much smaller base current. Amplifiers can be designed to produce as output an enlarged version of an input, whether of current, voltage or power, and at audio frequencies (20 to 20 k Hz) or at radio frequencies 20 k to 1000 MHz or more).

They are used in most electronic devices at some stage, as although transistors operate using very small power sources, the output of a loudspeaker or television tube is very much greater. Signals of various kinds need to be amplified to be of much practical use: for example, the radio or TV signals picked up by an aerial often need initial amplification before being decoded and converted into sound and vision with even more amplification.

The general symbol for an amplifier is:

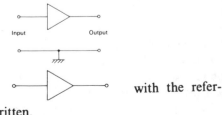

often abbreviated to just with the reference or zero level left unwritten.

The circuit of Fig. 8.17 is the simplest example of a voltage amplifier.

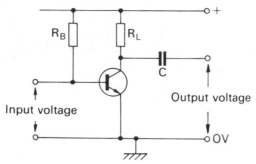

Fig. 8.17

If a small alternating voltage is applied as the input, a larger but *inverted* alternating voltage of the same frequency appears at the output. A cathode ray oscilloscope (Section 9.8) connected to the input and output terminals in turn would display the curves shown in Fig. 8.18. An amplification factor of about 70 can be obtained in this way. Note the resistors R_B and R_L similar to those in Figs 8.8 and 8.13. The capacitor C serves to make the output voltage symmetrical about the zero level.

Circuit diagrams for a simple temperature-stabilised amplifier and

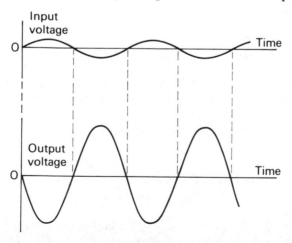

Fig. 8.18

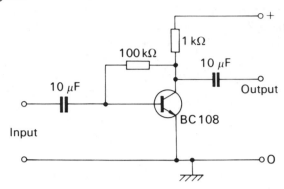

Fig. 8.19

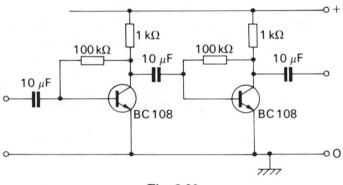

Fig. 8.20

a two-stage voltage amplifier are shown in Figs 8.19 and 8.20. An amplification factor of several thousand could be obtained with the latter circuit.

The output from the two-stage amplifier would not be inverted as in the single stage one.

8.11 Other semiconductor devices

Thermistor
Whilst metallic conductors generally increase their resistances when they are heated, the effect is not all that large – a doubling of resistance

for a temperature rise of, say, 300 degrees C. In the case of semiconductors (e.g. silicon), as we have seen, there is a very much larger effect but in the opposite sense – their resistances fall when they become hot, often by very large amounts. The thermistor TH3, for instance, has a resistance of 380 Ω at 25°C but only about 30 Ω at 100°C. Their main use is in electrical thermometers where changes in resistance can be made to show temperature changes directly, one practical advantage being that they can be made quite small and placed in awkward or dangerous positions well away from the rest of the circuitry and the operator. Because of their small size, very localised temperatures can be measured. However, special thermistors can also be made which give large *rises* of resistance with temperature, a property useful for protection of circuits against overheating.

Thyristor

This is a four layer p-n-p-n sandwich with three contacts: *anode*, *cathode* and *gate*. Under forward voltage it conducts when the gate is positive, and stays conducting on removal of the positive voltage until the anode voltage is removed or reversed. In the circuit of Fig. 8.21 the lamp will light when both S_1 and S_2 are closed, but will then remain on when S_2 is opened.

In AC circuits a thyristor can allow conduction at any positive voltage at which the gate is triggered. The effect is that the thyristor conducts only for part of the AC cycle, as in Fig. 8.22. The timing of the gate pulses determines the point at which conduction begins. By using a double 'back to back' thyristor, called a *triac*, conduction can be allowed on part of the negative half cycle as well (compare half and

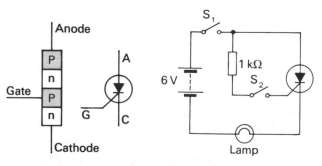

Fig. 8.21

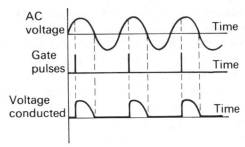

Fig. 8.22

full wave rectification – Section 7.51). This type of current control is far more efficient than using a variable resistor, since there is almost no power loss when the thyristor is off. Theatre lighting, for example, can be controlled using thyristors without the generation of large amounts of heat in the dimmers. The basic circuit for this is shown in Fig. 8.23, in which the variable resistor controls the proportion of the AC cycle during which conduction occurs. (A *diac* is a double Zener diode connected 'back to back'.)

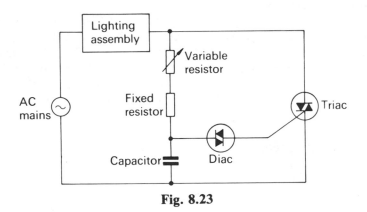

Fig. 8.23

Light-dependent resistor (LDR)
Sometimes called a photoconductive cell, this is a resistor made from cadmium sulphide whose resistance falls when illuminated externally. The ORP 12, for instance, has a resistance of about 10 MΩ in darkness but only 1 kΩ in daylight – a decrease by a factor of ten thousand. An application was illustrated in Section 8.8.

8.12 Preferred values

For the majority of circuits the exact values of components like resistors and capacitors are not critical, so it is not necessary for the manufacturing process to be very precise. It is enough as a rule to have a resistor, for example, with $\pm 20\%$ of its stated value, or for more precise purposes perhaps $\pm 10\%$ or even $\pm 5\%$. This means that only certain values need be made, provided that the tolerances either way overlap with one another. For example, at the $\pm 20\%$ level, a nominal $100\,\Omega$ resistor might have any value between 80 and 120 Ω, and one of $150\,\Omega$ any value between 120 and 180 Ω. Thus only 100 and $150\,\Omega$ resistors are needed at $\pm 20\%$ tolerance to cover the range 80 to $180\,\Omega$, and these nominal sizes are called *preferred values*.

Between 10 and 100 Ω, the preferred values at 20%, 10% and 5% tolerances are as follows:

20%	10			15			22			33			47			68		100
10%	10	12		15	18		22	27		33	39		47	56		68	82	100
5%	10 11 12 13	15	16 18 20 22		24 27 30 33		36 39 43 47		51 56 62 68		75 82 91		100					

Within each decade (multiples and sub-multiples of 10) only these particular sizes are needed, depending on the precision required – e.g. a 4·7 kΩ resistor, or 0·22 μF capacitor. Special value components can be obtained, of course, but the mass produced ones lie within the quoted levels of accuracy.

8.13 Resistor codes

Colour codes
The sizes of resistors are often indicated by a system of coloured bands. Three bands are used for the value of the resistance and a fourth for the tolerance according to the following code:

Number	Colour	Tolerance	Colour
0	black	5%	gold
1	brown		
2	red	10%	silver
3	orange		
4	yellow	20%	(no fourth band)
5	green		
6	blue		
7	violet		
8	grey		
9	white		

The first two bands give two figures for the resistance but the third indicates the number of noughts to be added, a system which fits well with the preferred values listed in the previous section. Fig. 8.24 shows a resistor with four bands.

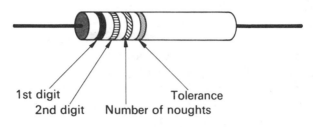

1st digit
2nd digit   Number of noughts   Tolerance

Fig. 8.24

Thus, for example, the sequence blue-grey-orange-silver would indicate a 68 000 Ω (or 68 kΩ) resistor on the $\pm 10\%$ series. A three-band sequence brown-black-red would identify a 1000 Ω (1 kΩ) resistor of 20% tolerance.

Printed codes

An alternative system of letters and figures is also in use. The letters R, K and M indicate 1, 1000 and 1000 000 respectively, with the position of the letter in relation to the figures indicating a decimal point. The following examples illustrate how the code works:

Value	10Ω	4·7Ω	1kΩ	3·3kΩ	680kΩ	2·2MΩ	0·15Ω
Code	10R	4R7	1K0	3K3	680K	2M2	R15

8.14 Diode and transistor codes

Two systems of codes are in use to identify the many different diodes and transistors available.

The American system uses an initial number (1, 2 or 3) followed by the letter N and a four-digit number. The initial number identifies the type of device: 1 = diode, 2 = transistor, 3 = thyristor. The letter N is common to all devices, and the last four digits form a registration number which has no technical significance. An example would be 2N2037.

The modern European (PRO ELECTRON) system is more informative and uses five characters: two letters followed by either three figures or a letter and two figures. In both cases the first letter shows the material from which the semiconductor is made (A = germanium, B = silicon, C = gallium arsenide) and the second letter indicates the most common application for which the device was designed (e.g. A = signal diode, C = audio-frequency low power transistor, E = tunnel diode, F = radio frequency amplifier, P = radiation sensitive device, S = low power switching transistor, Y = high power diode, Z = Zener diode). If the last three characters are all numbers, the device is suitable for entertainment or consumer equipment. If the last three characters are a letter followed by two numbers, the device is intended for industrial or professional equipment. Hence, for example, coding such as BC107 and BFY32. An older coding system prefixed semiconductor devices by 0 (meaning no heater voltage required), as in 0A81 and 0CP71 which denote a diode and a phototransistor respectively.

More complicated devices are coded differently, and integrated circuits are usually designated by four- or five-digit numbers. Manufacturers' catalogues and specialised publications often include the code numbers with other details of items on the market.

8.15 Integrated Circuits (IC)

The techniques of manufacturing components for electronic circuits now allow for thousands of circuit elements – resistors, diodes, transistors, capacitors – to be mounted, complete with inter-connections, on a silicon 'chip' about 5 mm square and 0·5 mm thick. The silicon has to be of a purity of about one part in ten thousand million, and is made in a cylindrical bar about 10 cm across. In manufacture, wafers are cut from the bar and between 200 and 300 identical circuits are formed alongside one another on each 0·5 mm wafer. After individual testing of each circuit, the 30% or so functional ones are mounted in a plastic case with gold wires connecting to the external pins. The whole IC would be about 2 cm long – Fig. 8.25. The number of external connecting pins varies between 8 and 24 depending on the particular circuits contained on the chip.

The advantages of an integrated circuit are its size, relative

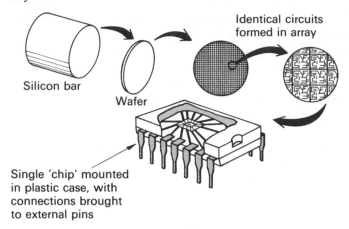

Identical circuits
formed in array

Silicon bar

Wafer

Single 'chip' mounted
in plastic case, with
connections brought
to external pins

Fig. 8.25

cheapness and reliability. Many different designs of IC are available, and practical electronics has become much more a matter of using ready made circuitry than laboriously building up circuits from separate components. One result is that, for many people, the main concern now is not so much about how ICs work but, rather, the uses to which they can be put. A *systems approach* is the way electronics is tackled through ICs and pre-manufactured circuitry instead of through detailed knowledge of the characteristics of individual components.

ICs form the basis of the development in computer design which has substantially reduced both costs and sizes (relative to power), and greatly extended the range of commercial, educational and domestic applications. There seems to be almost no limit to the ingenuity of the microelectronics industry in putting on to a silicon 'chip' virtually any required circuitry. No doubt further developments in electronics will make even those described here seem old fashioned, however, just as the radio valve has been made obsolete by the transistor, and the transistor by the integrated circuit. . . .

9

Atomic Bits and Pieces

9.1 Building bricks

Right at the beginning of this book the idea of atoms and molecules was introduced to describe the behaviour of matter. That idea has been one of the most fruitful in all science and evidence can be brought to suggest that it is at least a close approximation to the real thing (as far as we can tell the real thing!). The job in hand for this chapter is to review what we know about the next layer of smallness – the bits that make up an atom, the atomic particles.

9.2 Atomic structure – what does an atom look like?

If a powerful enough microscope were available we could answer that question, perhaps, but as it is we have to rely on second-hand evidence. The present view is that atoms contain three kinds of particles – **electrons, protons** and **neutrons**. The protons and neutrons are found together in the central part of the atom – the **nucleus** – and account for almost all the mass. Electrons by contrast are outside the nucleus, a long way from it compared to the diameter of the nucleus, and are much less massive than the other two particles. If a full stop represented roughly the size of a proton, an electron in the smallest atom would probably be some 100 metres from it!

The particles differ also in electric charge. Electrons carry a negative charge, protons an equal positive charge and neutrons no charge at all. The nearest visual image we can get is a small compact nucleus containing neutral and positive massive particles, with negative light particles relatively far away and the majority of the atom being empty space!

9.3 Evidence for the nucleus

At first glance the idea of atoms being largely full of nothing seems to be at variance with our experience of solid matter. A table top or hammer head does not feel anything but solid – yet there is strong evidence that these bulk properties are due to strong forces between atoms and molecules, but if a sharp enough probe is to hand solid matter does in fact behave like a sieve.

An important experiment was done in 1911 by Rutherford. Using gold as a target he fired at it some atomic bits, and observed some startling results. Gold was ideal for this job because it could be beaten into very thin leaf and its atoms are quite massive. The bullets were α-particles, positively charged and light compared with the gold atoms. A rough sketch of the idea is shown in Fig. 9.1. The counter could detect α-particles arriving at any angle from the gold foil and measure the number arriving every second. Of course the experiment had to be done in an evacuated chamber and the number recorded on the counter had to be corrected for those events it recorded with no α-particles at all hitting the gold foil.

As was to be expected, the counter registered most arrivals when in the straight line through position 1, and the count went down gradually as the angle increased through positions 2 and 3. What was quite unexpected though was that when the counter was taken round

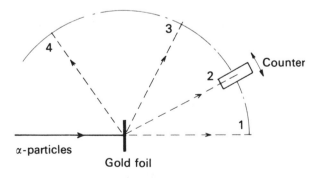

Fig. 9.1

to position 4 there still were more events recorded than could be accounted for by chance – in other words a few α-particles were being flung back by the foil through very large angles. Considering the thinness of the foil and the speed of the α-particles this was an almost incredible happening.

The interpretation was that the incident particles must have met a very concentrated repelling centre of force and, being positive themselves, that centre must itself be positive. It is only the 'head-on' collisions which give a large angle of scatter and since there are very few of them the atomic centres where the positive charge is concentrated must be very minute – Fig. 9.2. Nuclear diameters are something of the order of a millionth of a millionth of a millimetre, while atoms are about a hundred thousand times bigger than their nucleus.

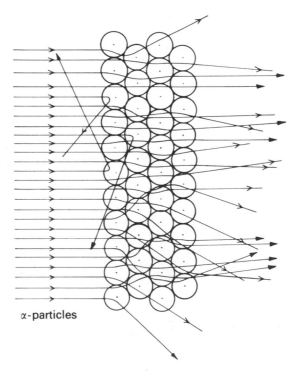

α-particles

Fig. 9.2

9.4 Electrons

A good deal has been written earlier in the book concerning the properties of electrons and what they can be made to do – especially in Chapter 7. Their own basic identifying features have not been dealt with though – their mass and charge in particular are what we need to know. In fact these are the quantities which enable us to identify *any* particle. For electrons the first quantity of any kind to be measured was a value for their charge to mass ratio, obtained in 1897 by Sir J. J. Thomson. This showed that electrons had either a much higher charge or a much lower mass than hydrogen ions. In the years 1909 to 1916 R. A. Millikan performed rather elaborate experiments to find the electronic charge directly.

9.5 Millikan's oil drop

The idea was to observe the motion of some very tiny oil drops falling freely under gravity at first, but later also under the influence of an electric field. If the drops carried a charge their motion was altered when the field was switched on and off. The charge on a particular drop was altered by directing X-rays on to it. In this way the charge was measured each time and when the numbers were worked out they were always multiples of a certain value, which was then assumed to be the charge on the electron. Fig. 9.3 shows a simplified sketch of the arrangement.

(It is rather like having a bag of marbles to weigh. Between each

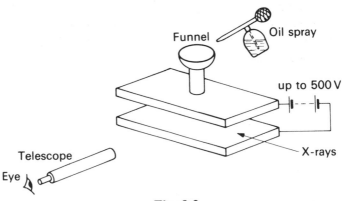

Fig. 9.3

weighing an unknown number of marbles either leave or enter the bag. If the bag is weighed often enough the difference each time will be a certain number of marbles and the HCF of these numbers will be the weight of one marble.)

Many experiments along similar lines have shown that there never occurs a charge smaller than this value. It is about 10^{-18} coulomb. To visualise this small number is very hard; 10^{18} electrons flowing past in 1 second give a current of 1 amp.

9.6 Deflecting electrons

The experiment of Millikan observes the oil drop's motion only when it is going at steady velocity, that is, when there is no net force on the drop. To get any information about the mass of the electrons we need to study them while they are accelerating since an object's mass determines what acceleration a given force will cause.

A convenient way of accelerating something without having to chase it over large distances is to make it go round in a circular arc. Such a path means an inwards acceleration towards the arc's centre, and by measuring the radius and speed the acceleration is known. Section 2.2 explored this problem.

Electrons can be dealt with in this way. A hot filament at one end of a vacuum tube produces plenty of electrons. They are speeded up by a positive voltage and then enter the region between two circular coils carrying an electric field. The coils make a magnetic field perpendicular to themselves and this bends the electron's path down or up into a circular arc. A fluorescent screen can show the path and enable its radius to be calculated. Figs 9.4 and 9.5 show the apparatus.

The arithmetic yields both the velocity of the electrons and the ratio of their charge to mass. In an experiment like this the electrons may

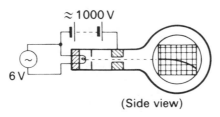

(Side view)

Fig. 9.4

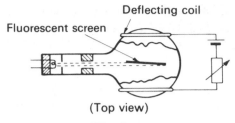

(Top view)

Fig. 9.5

get up to a tenth of the speed of light and the charge to mass ratio comes out at about 10^{11} coulombs per kilogram. This is a much more concentrated chunk of charge than we can put on any piece of metal and in fact electrons are little else but charge!

Combining a charge of 10^{-18} coulombs with 10^{11} coulombs per kilogram gives the electron's mass as 10^{-29} kilograms!!

9.7 Cathode rays – alias electrons

Originally electrons were 'discovered' in low pressure gas discharge tubes as unknown objects streaming off the cathode – or negative electrode – Fig. 9.6. The name 'cathode rays' has stuck for electrons, though now we do not produce them by the brute force method of very high voltage discharges.

The subtle, and much safer, way of getting electrons is to use the **thermionic** effect. A heated filament by itself will give electrons,

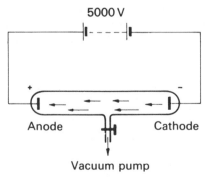

Fig. 9.6

'boiled off' from its surface; better still is to use a filament to heat another specially treated surface able to emit electrons copiously. This technique is used in radio valves and television tubes and in a device mentioned already in Sections 5.17 and 7.51 called a Cathode Ray Oscilloscope.

9.8 CRO – Cathode Ray Oscilloscope

A rough layout for a CRO is shown in Fig. 9.7. The cathode produces the electrons when heated by the filament. The anode system accelerates the electrons down the tube and focuses the beam to a narrow spot on the screen. Its position is shown by a phosphor inside the tube which emits light when electrons hit it. The anodes must be positive compared to the cathode, but the grid, held slightly negative, can control the intensity of the beam of electrons. All the connections, of course, are brought out to the end of the tube.

Having got a finely focused spot on the end of the tube the next thing to do is to deflect the spot. There are two pairs of electrodes between which the beam passes on its way down the tube. If potential differences are applied to these plates the electron beam will be deflected under the influence of the electric fields between the plates. The deflecting plates are usually termed 'X' and 'Y' according to which direction they cause the beam to move. Some simple examples are given in Fig. 9.8; one of each pair of plates is usually earthed. Used like this the device is a voltmeter – and one that takes no current to operate it – the ideal arrangement for a voltmeter.

The main virtue of a CRO is that since the electrons have so little mass they can react almost immediately to any *change* of voltage on

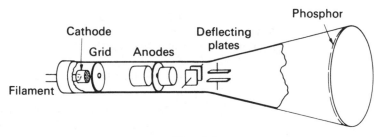

Fig. 9.7

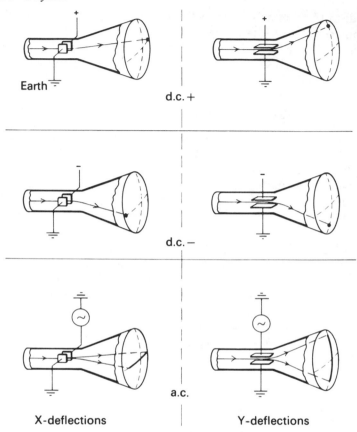

Earth

d.c. +

d.c. −

a.c.

X-deflections

Y-deflections

Fig. 9.8

the deflecting plates. CROs are used extensively to study *varying* voltages and with the addition of a circuit called a **time base** will draw a voltage-time graph. What the time base does is to drag the electron beam across the screen at a steady speed, and when it reaches the edge it flicks back to the beginning and repeats the trip. The voltage to be displayed is applied to the Y-plates and the time-base circuit to the X-plates, and if the two signals are synchronised a stationary trace is seen where the spot goes over the same path on each transit of the screen.

Any fluctuating effect which can be converted into a voltage variation can be displayed on a CRO. The easiest is the pressure changes concerned with a sound wave – a microphone will convert these into voltages and the CRO enables us to 'see' a sound wave – Section 5.17 referred to this.

All sorts of 'transducers' have been devised to convert pressure, temperature, magnetisation, current and other physical changes into voltage variation so that the oscilloscope has become a versatile instrument able to make a visual trace of almost *any* changing quantity.

9.9 Building up atoms

We have three particles from which to build up a sequence of atoms – electrons (e), neutrons (n) and protons (p), and we know the number of electrons and protons must be equal since bulk matter is neither positively nor negatively charged. Furthermore, the mass of the atom is made up by the number of neutrons and protons since electrons are so light they contribute hardly anything to the mass. All we need is a knowledge of relative atomic masses and we can build up a series of elements in increasing mass. Chemists have long had such information, the lightest element being hydrogen.

The simplest atom, hydrogen, needs only $1e$ and $1p$, and will therefore have a mass of 1 unit. Here are some common elements with their atomic structures.

Element and chemical symbol		Atomic structure			Units of mass	Nuclear symbol
Hydrogen	H	$1e$	$1p$		2	^1_1H
Helium	He	$2e$	$2p$	$2n$	4	^4_2He
Carbon	C	$6e$	$6p$	$6n$	12	$^{12}_6\text{C}$
Oxygen	O	$8e$	$8p$	$8n$	16	$^{16}_8\text{O}$
Neon	Ne	$10e$	$10p$	$12n$	22	$^{22}_{10}\text{Ne}$
Aluminium	Al	$13e$	$13p$	$14n$	27	$^{27}_{13}\text{Al}$
Calcium	Ca	$20e$	$20p$	$20n$	40	$^{40}_{20}\text{Ca}$
Silver	Ag	$47e$	$47p$	$61n$	108	$^{108}_{47}\text{Ag}$
Tin	Sn	$50e$	$50p$	$69n$	119	$^{119}_{50}\text{Sn}$
Lead	Pb	$82e$	$82p$	$125n$	207	$^{207}_{82}\text{Pb}$
Uranium	U	$92e$	$92p$	$146n$	238	$^{238}_{92}\text{U}$

The nuclear symbols in the last column show the number of protons, or **atomic number**, as the lower figures and the total number of neutrons and protons, or **mass number**, as the upper figure.

9.10 Isotopes

What is it that makes, say, carbon chemically the stuff we call carbon? Which atomic quantity identifies the element? When materials react chemically it is only the electrons which come in contact between molecules or atoms; so as long as there are six electrons the atom will be chemically the same as carbon. Six electrons means six protons too, so the identifying thing is the **atomic number**. Carbon always has $6e$ and $6p$, but what do the neutrons do? The simplest way to think of the neutrons is as binding agents for the protons which otherwise would fly apart because of their positive charges. In carbon, though, why should there be six neutrons too? Could there be seven or eight? If so we would have $^{13}_{6}C$ and $^{14}_{6}C$. Do these nuclei exist? Yes they do. They are still carbon, though not the same mass as the common variety and are called **isotopes** of carbon.

Nearly all elements have more than one isotope, though not all of them are stable. $^{14}_{6}C$, for example, disintegrates very slowly and can be used for estimating the age of fossilised materials.

9.11 Radioactivity

Many isotopes among all those theoretically possible are not stable. They show one or more kinds of **radioactive** decay which affect the nuclei of the atoms and ultimately they end up as one of the ordinary stable isotopes. The main characteristic of all kinds of radioactive decay processes is that the rate at which the nuclei change their composition depends on how many there are at the time. This means that the decay proceeds at a gradually reducing rate but never actually peters out altogether. The number of unaffected nuclei goes down by the same *fraction* in equal intervals of time. A period often quoted for a radioactive nucleus is its **half-life** – the time after which half of the atoms in a given sample have decayed.

If an isotope has 1 000 000 atoms at a certain time, one half-life later it will have $\frac{1}{2}$ million undecayed atoms, another half-life later only $\frac{1}{4}$ million, a third half-life later $\frac{1}{8}$ million, etc. For real radioisotopes the half-life might have almost any value from tens of thousands of millions of years to tiny fractions of a second.

For instance $\quad ^{238}_{92}U \rightarrow ^{234}_{90}Th + \alpha, 4\cdot5 \times 10^9$ years

whereas $\quad ^{214}_{84}Po \rightarrow ^{210}_{82}Pb + \alpha, 0\cdot000\ 16$ second.

9.12 The α-decay process

All mechanisms of decay involve the nucleus. Three distinct processes were discovered around 1900 and for want of a better set of names the particles emitted were called α, β and γ radiation. Once again these arbitrary names have stuck, though we know now just what is involved in each process.

α-particles are the nuclei of the helium atom, ^4_2He; that is, two protons and two neutrons. This is a particularly stable group of particles and there is some evidence that they exist in this grouping inside large nuclei. Nuclei made up of whole numbers of α-particles are also very stable, e.g. $^{16}_8O$, $^{12}_6C$, $^{40}_{20}Ca$.

A nuclei which is α-active shoots out an α-particle, Fig. 9.9. This means that the identity of the nucleus changes, since it loses two protons. Its mass also alters due to the four units in the α-particle, for example:

$$^{190}_{78}Pt \rightarrow ^4_2He + ^{186}_{76}Os$$

The 'daughter' nucleus is of a different chemical type from the 'parent'. (The two extra electrons left on the complete atom quickly leave the region of the nucleus due to its reduced attractive force for them.) This is the nearest we have come to achieving the old alchemist's dream of the transmutation of one element into another,

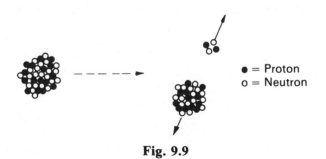

● = Proton
o = Neutron

Fig. 9.9

though it is not really under any sort of control and certainly will never change lead into gold!

Being rather heavy things as particles go, α-particles cause a great deal of chaos when they travel through ordinary matter and quickly lose their energy by collision. They knock electrons off atoms and leave a trail of atomic debris in their wake. It is rather like firing a cannon ball at a forest of saplings – great havoc for a short distance. The ions formed by successive collisions can be used to mark the track of the α-particle – Section 9.22.

α-particles cannot penetrate much more than a few cm of air and a single sheet of paper is enough to absorb them altogether. They present no physiological danger unless the source of the α-particles is either swallowed or inhaled.

9.13 The β-decay process

An alternative way in which a nucleus can spontaneously change itself is by emitting an electron or β-particle. At first sight this is not an expected occurrence since nuclei do not contain any electrons! The answer lies in the fact that the neutron and proton do not quite have the same mass. Protons as far as we know are stable particles, whether on their own or in the nucleus; neutrons on the other hand are stable inside nuclear matter for the most part, but free neutrons are radioactive with a half-life of some 11 minutes, giving a proton and an electron – Fig. 9.10.

$$n \rightarrow p + e$$

Looking again at the list of atomic arrangements in Section 9.9, it is clear that the heavier elements tend to be over-endowed with neutrons and hence more likely to be β-active. A nucleus in which a neutron undergoes the β-decay process will alter its identity; it loses a neutron

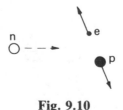

Fig. 9.10

but *gains* a proton, for example:

$$^{90}_{38}\text{Sr} \rightarrow e + ^{90}_{39}\text{Y}$$

Electrons are much lighter and smaller than α-particles and so they do not cause so much ionisation as they travel through matter. They suffer many collisions which deflect them, but even though they may start off at a very high speed they get through only a few metres of air before being stopped. It is like firing a pellet at a forest of saplings, quite a lot of trees are hit and the pellet bounces off them to penetrate quite a long way.

High energy β-particles can cause damage to human tissue, but even these can be stopped by a cm or so of aluminium and they are not the main danger from nuclear explosions.

9.14 The γ-decay process

The names α, β and γ were assigned to the three main radioactive reactions before it was known just what was involved in them. α and β processes involve particles, so we talk about α-particles and β-particles (sometimes loosely β-rays), but the γ reaction does not involve a particle like the other two. In this decay process the nucleus simply shakes itself down from one energy state to a lower one and the surplus energy appears as radiation, light waves of very high frequency and very short wavelength, $\approx 10^{-12}$ mm.

γ-decay often occurs after one of the others. Frequently the daughter nucleus formed after α or β emission is not in its lowest energy state and γ emission follows, for example:

$$^{32}_{15}\text{P} \rightarrow ^{32}_{16}\text{S} + \beta, \qquad \text{then} \quad ^{32}_{16}\text{S} \rightarrow ^{32}_{16}\text{S} + \gamma$$

or $$^{236}_{92}\text{U} \rightarrow ^{232}_{90}\text{Th} + \alpha, \qquad \text{then} \, ^{232}_{90}\text{Th} \rightarrow ^{232}_{90}\text{Th} + \gamma$$

The nucleus involved with γ-decay does not change its chemical name, unlike the results of α- or β-decay. The radiation is extremely penetrating and this is the one which presents the grave danger as a result of nuclear explosions. Many cm of lead or even metres of concrete are required adequately to shield a person from the effects of γ-radiation. Genetic changes are caused on the chromosomes by γ-rays and usually these are not beneficial to say the least.

9.15 Natural radioactivity

Quite a few isotopes are spontaneously radioactive. For these still to be present in the earth they must have a half-life which is comparable with the age of the earth. If the half-life is appreciably less than this any such substances would long ago have decayed to unobservably small amounts. This is the reason why the chemists' list of naturally occurring elements stops at uranium, atomic number 92. The heavier elements are spontaneously radioactive with quite short lifetimes and so do not occur naturally in the earth. They can be made in the laboratory up to about atomic number 105 by artificial means but are not stable. For instance, plutonium 242 decays by α emission to uranium 238:

$$^{242}_{94}\text{Pu} \rightarrow {}^{238}_{92}\text{U} + \alpha, \text{ half-life } 380\,000 \text{ years.}$$

(Recently some interesting evidence may have been found in the particles arriving at the earth's atmosphere from outer space, cosmic rays, to indicate that atomic number 110 or thereabouts would in fact be stable. It remains to be seen whether any will be discovered anywhere on earth, though really there is no reason why the earth should have been equipped originally with every possible stable isotope!)

Some isotopes close to the heavy end of the list are also naturally radioactive, often by several alternative routes, but generally ending up as one of the isotopes of lead, for example:

$$^{238}_{92}\text{U} \rightarrow {}^{206}_{82}\text{Pb} +$$
$$\text{several } \alpha, \beta \text{ and } \gamma\text{s, half-life } 4\cdot5 \times 10^9 \text{ years}$$

and $^{232}_{90}\text{Th} \rightarrow {}^{208}_{82}\text{Pb} +$
$$\text{several } \alpha, \beta \text{ and } \gamma\text{s, half-life } 1\cdot4 \times 10^{10} \text{ years.}$$

9.16 Artificial radioactivity

By bombarding ordinary isotopes with one kind of particle or another it is often possible to obtain an isotope which has a conveniently short (or long) half-life for a useful application. The change might happen naturally, for instance, as in the case of $^{14}_6\text{C}$:

$$^{14}_7\text{N} + \text{neutron} \rightarrow {}^{14}_6\text{C} + \text{proton.}$$

The 'radio carbon' ^{14}C decays with a long half-life, 5600 years, and it can be used to date the time of laying down of fossils and vegetable matter, since the proportion of ^{14}C in living material is known and constant.

Ordinary magnesium ^{24}Mg can be changed to sodium by a reaction involving neutrons:

$$^{24}_{12}\text{Mg} + \text{neutron} \rightarrow {}^{24}_{11}\text{Na} + \text{proton}.$$

The ^{24}Na isotope is β-active back to magnesium with a half-life of around 15 hours:

$$^{24}_{11}\text{Na} \rightarrow {}^{24}_{12}\text{Mg} + \text{electron}.$$

This reaction is considered again in Section 9.19.

9.17 Atomic energy – Fission

One of the isotopes of uranium, ^{235}U, will split into two nearly equal parts if a fairly slow neutron hits it. The reaction also produces extra neutrons and a considerable amount of energy. One possibility is:

$$^{235}_{92}\text{U} + \text{slow neutron} \rightarrow$$
$$^{144}_{56}\text{Ba} + {}^{90}_{36}\text{Kr} + 2 \text{ fast neutrons} + \text{energy}.$$

This splitting process is called **nuclear fission** and it offers two main possibilities as an energy source:

(a) Uncontrolled – a bomb!

In a large lump of uranium the two fast neutrons made after one fission may be slowed down by collisions until they are slow enough to initiate two more fissions. Four more neutrons can then do the same and the result is an uncontrolled chain reaction quickly releasing huge amounts of energy in a very short time – an atomic bomb, in fact. Heat and blast are the main destroyers, but the fission products are often themselves radioactive and radiation sickness can be caused years after such an explosion, as happened at the end of the Second World War in 1945.

(b) Controlled – an atomic pile

If the uranium is kept in small pieces and if some of the neutrons produced escape altogether or get absorbed by suitable materials, the reaction rate can be kept under control and even made adjustable.

This is the basis of atomic power stations where the steam is generated from heat produced by an array of 'fizzing' uranium rods. The workers there are protected from radiation by walls several feet thick and the waste materials are disposed of down deep disused mine shafts or far out to sea.

9.18 Atomic energy – Fusion

The opposite process, building up a larger nucleus from smaller bits, can release even larger quantities of energy. Several *fusion* reactions, as they are called, are possible, for example:

$$_1^1H + _1^1H \rightarrow _1^2H + _1^0e + energy$$
$$_1^2H + _1^2H \rightarrow _2^3He + _0^1n + energy.$$

The snag with these reactions – basically the kind which keep the sun going – is that they need temperatures in millions of degrees to get them started. So far only the uncontrolled version has been developed, called the **hydrogen bomb**, and the necessary high temperature has been achieved by detonating it with an atomic (fission) bomb!

The problem of controlling this ultimate source of energy is clearly formidable – how to achieve safely the millions of degrees and what to contain it in? The answers have so far eluded scientists, but success will be worth striving for since the sea can provide us with ample deuterium to use as a fuel – it would be the complete answer to the world's energy supply problems.

9.19 Radioactivity put to work

The United Kingdom Atomic Energy Authority provides a concrete example of the usefulness of radioactive technology. The electrical energy results from the heat released during the controlled fission of uranium atoms. Nuclear power stations are being built all over the world as further evidence that this type of power generation is a commercial proposition.

Radioactive isotopes of almost any element can be made by using the radiation available inside atomic piles or from the many particle accelerators now in action. The unstable isotopes are put to a great variety of uses, some medicinal or biological, others in heavy or light industry.

$^2_1$H can be used in organic chemistry to trace the path of hydrogen atoms in the formation of complex molecular substances.

$^{47}_{20}$Ca, given to chickens, can help to solve the problem of too soft shells on eggs.

$^{32}_{15}$P can be taken in food and used to trace the processes involved in tooth growth and decay.

$^{59}_{26}$Fe can be used in a rolling mill of a steel works to check the constant thickness of steel sheet.

$^{24}_{11}$Na, taken in the form of salt, can be used as a signpost to the stages of digestion inside a human body.

The toothpaste industry uses radioactive sources as an automatic check on the fullness of toothpaste tubes on a mass production line.

Flaws inside metal castings can be detected by the way they scatter radioactive particles fired into them.

9.20 X-rays

The atom also provides another weapon which has made a great impact on many branches of science – X-rays. These were originally discovered somewhat accidentally by Röntgen in 1895 and he dubbed them X-the-unknown. Instead of broadcasting the news immediately he made the best of his good fortune and performed all the obvious experiments and tests before publishing the information. Within a very few years these penetrating rays were in use in the exploration of bone fractures.

X-rays are produced when electrons are rapidly slowed down. In practice this means firing a beam of electrons at a heavy target, often tungsten. The radiation comes partly from the decelerating electrons and partly from the target atoms. A lot of heat is generated too and the target may have to be embedded in a good conductor like copper or even water-cooled! Fig. 9.11 shows a simple X-ray tube.

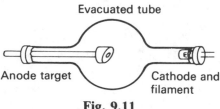

Evacuated tube

Anode target   Cathode and filament

Fig. 9.11

X-rays are the same kind of thing as γ-rays, but some hundred thousand times longer in wavelength. Their wavelength is a measure of how penetrating they are: 10^{-7} mm is a typical value compared with about 10^{-4} mm for visible light.

9.21 The complete spectrum

γ-rays and X-rays form one end of a much larger range of radiation which Clerk Maxwell showed was a combined effect of electric and magnetic fields. We use the general term **electromagnetic radiation** for all waves of this kind, and the special feature of γ- and X-rays is that they fall at the short wavelength end of the range. Visible light forms only a small part of this 'spectrum' and happens conveniently to be able to penetrate the earth's atmosphere and give us the sensation of sight when it reaches the retina of the eye.

Other radiations of this same kind produce a feeling of warmth (infra-red) or cause sunburn (ultra-violet) or are used for radar or radio communications. The complete set of wavelengths covers an enormous range from about 2 km to 10^{-12} mm. Since all the different kinds of wave in Fig. 9.12 are essentially the same they will travel at the same speed through space, so we could equally well specify them by their frequencies. Multiplying frequency by wavelength must always give the speed of the waves (Section 5.1) which is 3×10^8 metres per second.

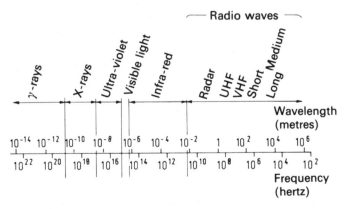

Fig. 9.12

9.22 Seeing atomic particles

The problem of seeing molecules was discussed in Section 1.2 and proved difficult enough. Large molecules *can* just about be photographed using special techniques and electron microscopes, but when we wish to make the problem about a million times smaller there is no hope of a direct picture at all. Protons and neutrons have diameters of about 10^{-12} mm, the smallest atom is 10^{-7} mm across, but electrons are almost immeasurably small even in this world of tiny distances. Such techniques as are available offer only second-hand evidence of where atomic particles have been.

The trade mark which many moving particles leave behind them is a trail of ionisation. The path taken by an atomic particle, especially a charged one, is littered with ions, debris from the material the particle has ploughed its way through, and these form the basis for several methods by which the path can be made visible.

9.23 Cloud chambers

A steamy bathroom gives some indication of how a cloud chamber works. When the volume of air in the room becomes overloaded with water, vapour condensation occurs on the windows or walls, which are generally slightly cooler than the room as a whole. In a bathroom there is no hindrance to condensation, but if the walls were smooth and the air perfectly clean it would be possible for the water vapour to remain uncondensed and the air 'supersaturated'. Any disturbance then, such as the arrival of a speck of dust, can initiate the condensation and a droplet forms around the speck of dust.

Cloud chambers use a technique similar to this with a more volatile liquid than water producing drops of liquid along the paths taken by any charged particles. A simple version of a cloud chamber can easily be made – Fig. 9.13 – provided some solid carbon dioxide (dry ice), or liquid air, is available. The trails of droplets have to be photographed stereoscopically and it is possible to identify the particles from the thickness, range and curvature of the tracks. This last effect is obtained if the chamber is put inside a magnetic field of known strength.

Fig. 9.14 shows what the tracks of α-particles look like; they are quite short but show heavy ionisation. The particles quickly dissipate their energy.

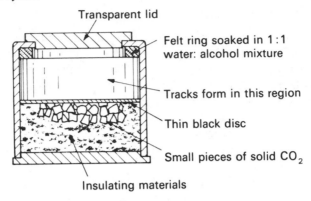

Transparent lid

Felt ring soaked in 1:1 water: alcohol mixture

Tracks form in this region

Thin black disc

Small pieces of solid CO_2

Insulating materials

Fig. 9.13

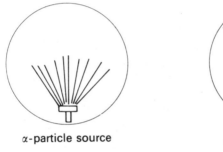

α-particle source

Fig. 9.14

β-particle source

Fig. 9.15

β-particle tracks are sketched in Fig. 9.15, much thinner than those of α-particles, due to significantly less ionisation. Being very much smaller and lighter particles the βs have a longer range before collisions bring them to a halt.

Uncharged particles leave no track in a cloud chamber, but often their presence can be inferred from examining tracks which originate apparently out of nothing. Fig. 9.16 is a drawing based on a photograph which indicates the presence of a neutron along path AB. The nitrogen nucleus it strikes recoils off to the left and another particle is ejected towards the right.

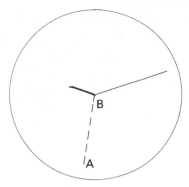

A nuclear collision

Fig. 9.16

9.24 Geiger–Müller tube

The ionisation caused by fast-moving atomic particles can be used in a quite different way to show their presence. If the ionisation occurs in a region of strong electric field which is already on the verge of sparking, a discharge can be triggered off by it and each spark effectively shows the arrival of a single particle.

This arrangement can be contained within a fairly small space and a Geiger–Müller tube is such a device. A necessary piece of auxiliary equipment is something to count the discharges separately (a scaler) or else to show the rate of arrival of pulses from the tube (a ratemeter). Either can be used quantitatively to measure or monitor radioactive materials: Fig. 9.17 shows the usual arrangement.

Geiger–Müller tubes have two special properties which must be

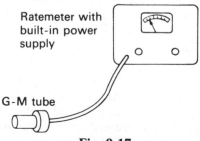

Ratemeter with
built-in power
supply

G-M tube

Fig. 9.17

remembered when they are used. The first is that they need a definite operating voltage, normally about 420 volts, and values much higher or lower than this cause the tube to be unreliable or even cause damage to it. The second property to remember is that there is a 'dead' time immediately after a pulse has been triggered, when a particle cannot give a pulse even if one arrives. This means there is a limit to the number of particles per second which can be counted reliably.

A further practical point is that Geiger–Müller tubes record events even when there is no radioactive material nearby. These counts are due to particles always present in the atmosphere and the 'background' count, as it is called, must be subtracted from any readings taken for a radioactive source. The particles are produced by the arrival from space of cosmic rays – mainly protons – which collide with the upper layers of the atmosphere to give a variety of particles at ground level. Depending on the location there might be some 30 or so registered by a Geiger–Müller tube every minute.

Geiger–Müller tubes can be made sensitive to α, β and γ radiations and even to X-rays.

9.25 Other techniques

A variety of other techniques has been developed for detecting and identifying fast moving particles which can be mentioned here. Special nuclear emulsions can be made for photographic plates which record the tracks of particles in such a way that they can be processed chemically to give a photograph directly. These are often used for high altitude investigations of cosmic rays, being usually sent up by balloons or rockets.

Certain particles cause a tiny flash of light when they collide with the right material and scintillation counters can be made which amplify and record the light flashes, enabling the particles to be detected. Solid state detectors work in a similar way, but here the arriving particle causes an electron to shift its state inside a semiconducting material and amplifiers make the change easy to record.

Bubble chambers work in the reverse way to cloud chambers. Instead of the particle giving rise to a trail of liquid droplets in a supersaturated vapour, they cause a trail of vapour bubbles in a superheated liquid. Using a liquid of known composition, like liquid hydrogen, any collisions can be used to identify the particles.

Index